The Shropshire & Montgomeryshire Railway

£8.95

The Shropshire & Montgomeryshire Railway

Eric S Tonks   M.Sc., A.R.I.C.

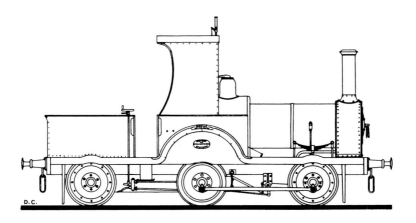

GAZELLE — *Drawing by Douglas Clayton.*

The Industrial Railway Society

First published by the author in 1949
Revised and enlarged edition published by the IRS in 1972
1972 edition reprinted in 2007 ©

Published by   Industrial Railway Society
                    24 Dulverton Road,
                    Melton Mowbray,
                    Leicestershire,
                    LE13 OSF
Email   sales@irsociety.co.uk
Web site   www.irsociety.co.uk
For details of Society membership send sae to:
                    Mr. B. Mettam, 27 Glenfield Crescent, Newbold, Chesterfield, S41 8SF

                    ISBN 978 1 901556 50 6

NOTES TO THE 2007 REPRINT
We are pleased to reproduce in full the text of the definitive 1972 edition of this book, edited by the late
Eric S. Tonks, former President of the Industrial Railway Society.

The only change to the 1972 edition being this page and re-designed covers.

Produced by Print-Rite, Witney, Oxfordshire. 01993 881662

CONTENTS

Acknowledgments

The S. & M. can well claim to have attracted more enthusiasts than any other light railway in the kingdom and the extent of our knowledge of this line is attributable in no small measure to the observations of these unknown "fans" — in the form of notes in all sorts of railway periodicals or sometimes passed on to others of like sentiments; my appreciation of their efforts is in no wise diminished by their anonymity.

I owe a special debt of gratitude to Mr. R. D. Chamberlin, whose ciné film of the line in its more romantic days first aroused my interest many years ago ; to Mr. E. T. Sloane, editor of the *Shrewsbury Chronicle*, whose kindness in placing at my disposal the files of the paper over a very long period has helped me so much with past history; to Mr. C. H. Calder, M.B.E., the Operating Officer, for generous facilities for inspecting the line under W.D. ownership; to various personal friends and correspondents — Messrs. N. J. Allcock, W. Beckerlegge, R. K. Cope, A. M. Davies, A. L. F. Fuller, E. Gibbons, H. Gray, T. C. Hancox, E. W. Hannan, P. Hindley, P. Laws, E. J. Lees, A. P. Miall, F. L. Pugh, M. Rhodes, B. Roberts, R. H. Stuart, D. Webb, R. P. White, W. K. Williams — for all sorts of data; to my wife for help with the proofreading and the scouring of dusty newspaper files; to the photographers, with special thanks to the proprietors, Locomotive & General Railway Photographs, and to Mr. E. C. Griffiths who generously made a photographic survey of the railway remains in 1970; and to Messrs. R. W. Clark and R. E. West for the maps and plans. Especially I must thank the late Mr. T. R. Perkins, who without doubt had a more intimate knowledge of this line than any other railway historian; he very painstakingly checked the original script from his own personal knowledge and notes and generously placed his collection of photographs at my disposal; without his help, the book would have been much the poorer.

Finally, my thanks to the Industrial Railway Society for publishing this revised edition, which includes information that has come to light since the publication of the first edition of 1949, and brings the story up to date — and, indeed, to a close.

87, Sunnymead Road,   ERIC S. TONKS.
Birmingham.
November 1971

The Potteries, Shrewsbury & North Wales Railway
Period 1861 - 1880

Promotion and Construction of the Railway

SHREWSBURY is a "county town" in the traditional sense, that is, the recognised centre and market for the agricultural country which still constitutes the major portion of Shropshire; and the plain of the Severn to the west of Shrewsbury has perhaps the richest pasture of any in the county. It is a land of scattered farmsteads and tiny villages, and certain areas are subject to periodic inundation by the river; and in consequence it had little to tempt the speculators of the period of the "Railway Mania." But Shrewsbury is also the gateway to Wales and in the middle of the last century many schemes were proposed for railways with Shrewsbury as the point of departure for the Principality; most of these never progressed beyond the paper stage and by 1865 Shrewsbury, from the railway point of view, was well and truly in the hands of the Great Western and London & North Western Railways, a factor which was to have profound influence on the future course of events; for whilst these two great companies were competitive in other spheres, their joint interest in Shrewsbury enabled them to wield powerful influence towards the suppression of any possible threat to their monopoly.

The various routes under their control comprised the Chester and Severn Valley lines owned by the G.W.R: the Crewe line owned by the L. & N.W.R: and the joint lines to Wellington and to Hereford. To the latter category was added on June 1st, 1861 the Rea Valley line, which ran from Sutton Bridge Junction on the Shrewsbury and Hereford to Minsterley, the only portion constructed of a much more ambitious scheme to connect Shrewsbury with Aberystwyth via Minsterley, Montgomery, Newtown and Llanidloes; this was later modified to run via Criggion and Welshpool, Newtown and Machynlleth, but never came to fruition. The last of the lines in the district to come under joint ownership was the connection from Cruckmeole Junction (Shrewsbury) to Buttington Junction on the till then isolated Oswestry & Newtown Railway, which had been opened officially as far as Welshpool on August 14th, 1860; the Shrewsbury-Buttington line was itself opened for traffic on January 27th, 1862, and passed into the hands of the G.W.R. and L.N.W.R. in 1865, when they obtained running powers into Welshpool.

It was into this hive of main line activity that the predecessors of the S. & M.R. timidly tried to thrust their way and (to continue the metaphor) were badly stung for their temerity. The first concrete proposal was the publication in the autumn of 1860 of the prospectus of the West Midland, Shrewsbury & Coast of Wales Railway, an optimistic project having as its ultimate aim the opening of an alternative route to Ireland. Starting from Shrewsbury, the railway was to run via Kinnerley to Porthywaen, up the Tanat Valley to Llangynog, thence to Llandrillo by means of a tunnel between a mile and a half and two miles long under the Berwyns, and onwards to Bala and Portmadoc; a total distance of 90 miles. Eventually it was hoped to reach that mirage-Mecca of North Wales, Porthdinlleyn (on the Caernarvon coast near Nevin) which featured in a number of schemes in the role of a potential rival to Holyhead as a point of departure for Ireland.

Having regard to the sparseness of the population along the route, and to the more than considerable engineering works which would be necessitated, it is hardly surprising that there was little local enthusiasm for the project; a traveller gazing at the frowning Berwyns from Llangynog platform might be forgiven for thinking the idea of a railway beyond that terminus as over-optimistic, to say the least. It was in the question of local support that this Company differed from its successors, for whatever the ultimate fate of the latter, they at least always managed — bitter experience notwithstanding — an enthusiastic send-off. This factor was to reveal itself almost at once; the good people of Llanfyllin, which the proposed line hoped, by deviation or branch, to serve, expressed their preference for being at the terminus of a branch line from Llanymynech on the Oswestry & Newtown rather than as an intermediate stop on a hypothetical cross country route; and this in spite of the presence at the meeting of Mr. A. C. Sherriff, the Manager of the West Midland, etc. Rly. In 1861, the Bill for the Railway was rejected by Parliament on Standing Orders, and as such was heard of no more.

The next attempt to build a railway across the Severn Plain was a more modest affair, and was authorized in 1862 under the simple title of the West Shropshire Mineral Railway; the chief promoter was Richard Samuel France, the proprietor of limestone quarries on Llanymynech Hill, and the avowed object of the railway was to facilitate the transport of stone from these and other quarries in the vicinity; it seems likely, however, from subsequent events and from the testimony of contemporaries, that Mr. France even then had more ambitious plans in mind.

In its original form, this line was designed to run from Llanymynech to Kinnerley, thence to a junction with the Shrewsbury-Welshpool line at Westbury, eleven miles from Shrewsbury; in 1863 and 1864, however, Parliamentary powers were sought in four Bills to modify the provisions and to change the title to the more imposing "Shrewsbury & North Wales Railway." Under this scheme, the junction with the Welshpool line was altered to

Red Hill, about two and a half miles from Shrewsbury, and running powers were to be obtained into the General Station, as it was now intended to carry passengers; two further sections of line were also proposed — a three mile extension from Llanymynech to Llanyblodwel to serve the Nantmawr lime quarries, and a six-mile branch from Kinnerley to Criggion granite quarries. Parliamentary consent was accorded all these proposals with the exception of running powers over the joint G.W.R.—L.N.W.R. line to General Station,* at the behest of the owning Companies—a premonitory rumbling of Jovian wrath which the S. & N.W.R. might have done well to ponder. Undeterred by this last difficulty, the Company built instead an independent line into Shrewsbury, parallel to the Welshpool line most of the way, finally crossing over the Hereford and Severn Valley lines and dropping down steeply to a terminus abutting the London road at Abbey Foregate; from the summit of this bank, a half-mile (or rather more) spur ran to a junction with the Shrewsbury-Wellington Railway. In 1865, a scheme to build a 14-mile branch from Llanymynech to Llanfair Caereinion along the Meifod Valley was also put before Parliament, but the powers were allowed to lapse.

Commencing from the quarries at Llanymynech, construction proceeded steadily along the whole of the 28 route miles authorized, the contractual work being undertaken by Mr. R. S. France and under the personal supervision of Mr. W. H. France; meanwhile, further advances in the aims and aspirations of the Company, involving the adoption of yet another and grander title, were under way. This came about by amalgamation with another struggling concern, the Shrewsbury & Potteries Junction Railway, promoted at the same period to connect Shrewsbury with Stoke on Trent by means of a line running via Market Drayton; the latter were authorized in 1865 to link the Wellington-Market Drayton and Shrewsbury-Crewe lines, but they now modified their plans to connect with the S. & N. W. line instead and in 1866 the combined undertaking, under the style of the "Potteries, Shrewsbury & North Wales Railway" sought Parliamentary sanction for this latter project and to revive the first stage of the old W.M.S. & C.W.R. scheme for an extension from Llanyblodwel to Llangynog and Portmadoc. This time, a great deal of public interest and sympathy was aroused, both in Shrewsbury (where an alternative outlet to industrial North Staffs was especially favoured) and elsewhere along the proposed route; and eventually, despite strenuous opposition from the Great Western and London & North Western Companies, Parliamentary approval was duly obtained; though, as events proved, the Company might have saved itself the trouble and expense of this additional legislation.

The line authorized under the powers conferred in 1865—now officially

---

* Earthworks suggesting a connecting line can still be seen close to the joint line at Red Hill, however.

referred to as the "Potteries, Shrewsbury & North Wales Railway; North Wales Section" —was completed in the summer of 1866 and, the Board of Trade Inspector having expressed his complete approval, the Shrewsbury-Llanymynech line was opened for public passenger traffic on Monday, 13th August, 1866; on the Criggion (or Breidden, as it was then termed) branch and the Llanyblodwel extension goods traffic only was conveyed. No ceremony attended the opening, no inaugural run, no speeches; merely a printed notice and a brief mention in the local press — a modesty becoming the Railway's accomplishments and excelled only by that attending its retirement.

Layout and Equipment

The course of the line constituting the "North Wales Section" of the P.S. & N.W.R. did not differ in essentials from the S. & M.R. of later years apart from the then possession of the Llanyblodwel extension. The terminus at Shrewsbury lies on the side of the road opposite to the Abbey Church, and indeed was laid on the site of the Abbey Refectory, the pulpit of which is still standing in a railed-off portion of the station yard; it was built probably in the time of Nicholas Stevens, Abbot from 1361-1399, and is scheduled as an ancient monument.

There was a very stiff climb from the terminus (the only gradient of any consequence on the railway), at the summit of which the line was joined on the left by the spur to the Wellington line, crossed the "Rea brook" and the Shrewsbury-Hartlebury and Shrewsbury-Hereford railways by girder bridges, and then descended to the level of the Shrewsbury-Welshpool line, with which it ran parallel to Red Hill, a distance of about two miles; in spite of this, there was no physical connection between the two, all exchange traffic being transferred over the spur. Beyond Red Hill, the line curved gently and was embanked through a spinney to cross the joint line by a girder bridge; thence to Kinnerley the route took an almost north-westerly course, apart from a small deviation at Shrawardine, where the Severn was crossed. Between Red Hill and Shrawardine a considerable number of earthworks had been necessitated, including two long cuttings and the rest of the distance mostly embanked. The viaduct at Shrawardine was the principal engineering feature of the railway and was a substantial structure of six spans, 320 feet in total length, constructed of wrought iron girders supported by cast iron and concrete pillars. To the west of the viaduct the railway was practically level all the way to Llanymynech, with very little earthwork but a multiplicity of level crossings, all of which were protected by gates.

Kinnerley, the junction of the Breidden branch, was the principal intermediate station, though not so important as in later days, when it was to become the railway's headquarters. The main line continued almost perfectly

12

straight for about four miles in a direction W.N.W. and then curved into Llanymynech station alongside the Cambrian platforms; the double junction at the southern end of the station was unique amongst light railway junctions. For a few hundred yards beyond, the P.S. & N.W.R. exercised running powers over the Cambrian lines and the Llanyblodwel extension then curved away sharply to the west and ran round the foot of Llanymynech Hill to the quarries on the western side. There was no feature of especial interest on this section apart from the double bridge carrying the Llanfyllin road and the Shropshire Union canal over the railway at Wern.

The Breidden branch, which can best be described as an S-shape

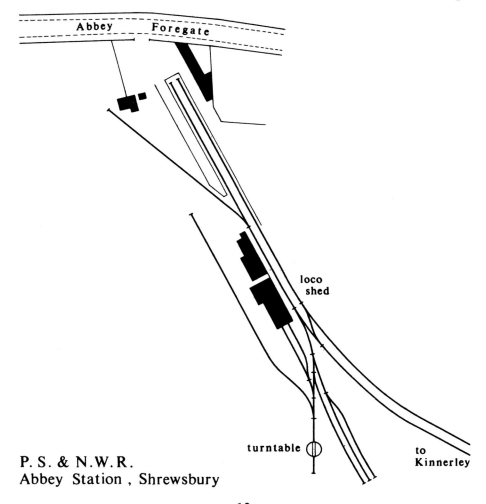

Abbey   Foregate

loco
shed

turntable

to
Kinnerley

P. S. & N.W.R.
Abbey Station , Shrewsbury

extending first south from Kinnerley and then west, was also level and plentifully blessed with level crossings; the Severn was again crossed, at Melverley, but in contradistinction to the viaduct at Shrawardine, which was a very sound piece of engineering, that at Melverley was a rather shaky timber structure which did not long survive the railway. At Melverley station (about ¾ of a mile north of the viaduct) the bridge carrying the adjoining minor road over the railway was an interesting seven-arched structure— unusually long for a rural branch line — of red and blue brick. The terminus at Criggion lay right at the foot of the richly-wooded Breidden Hills, the granite quarries on which provided the raison d'etre of the branch.

The remaining piece of line, the Abbey Foregate spur, must have been the most expensive half mile of all, comprising a high embankment — spanning the two branches of the Rea with a double arch brick bridge in one case and a girder bridge in the other — and a deep cutting including a bridge under the London road and another under a minor road before the joint line was reached.

The bridges, apart from those which have been mentioned, were not in general remarkable, though numerous enough. Most of those under the line were of the girder type, and the overbridges constructed of brick, but some had wooden arches on brick abutments; there were many of the former spanning small streams, especially between Kinnerley and Llanymynech along the valley of the Vyrnwy, and others over low lying ground liable to periodic flooding.

The 18-mile main line from Shrewsbury to Llanymynech was laid with double track, though this state of affairs was of short duration, one track — in some places the up, and in others the down — being removed (except for passing places in certain stations) in 1867; the unique double junction between the "Potteries" and the Cambrian Railways at Llanymynech remained, however. The Llanyblodwel extension and Criggion branch were both single track only. Chaired rails, 70 lbs. per yard, were used throughout and on the Criggion branch and a few odd sidings much of the original equipment survived to the end of the S. & M.R.

The railway boasted fourteen stations, including the four on the Breidden branch. Those on the main line were as follows, the distances from Shrewsbury being given in brackets: Red Hill (3), Hanwood Road (4), Ford (7½), Shrawardine (9½), Nesscliff (11¾), Kinnerley (13½), Maesbrook (16), Llanymynech (18) and Llanyblodwel (21). All but the last were provided with double platforms at first, but when the track was singled the one platform was removed from those stations not provided with passing loops, viz., Red Hill, Hanwood Road, Nesscliff and Maesbrook. The branch line stations, with distances from Kinnerley, were Melverley (2½), Crew Green (3½), Llandrinio Road (5) and Criggion (6).

Four stations—Shrewsbury, Nesscliff, Kinnerley and Melverley—were provided with brick buildings, the rest only having simple wooden structures usually comprising a booking office and a waiting room. As the head office of the railway, Abbey station at Shrewsbury was built on a (comparatively) more pretentious scale and sported a verandah and a covered approach to the platform; a small goods yard adjoined the station premises, together with a locomotive shed, repair shops, turntable and sundry offices. There was also a siding to the works of the Midland Wagon Company. Llanymynech, which for practical purposes could be regarded as the other terminus of the main line, had a wooden building (with verandah !) on one platform only, a small engine shed with a turntable, and a water tank mounted on timbers, for the refreshment of the locomotives. Kinnerley had a bay platform for branch trains and most of the stations were provided with one or sometimes two sidings (e.g. Ford, the name of which was subsequently altered to Crossgates); that at Hanwood Road descended very steeply to the roadside below the platform. There were, naturally, extensions at Llanyblodwel and Criggion to the quarries.

A map published in 1874 by Edwin Slater of Manchester for the Wynnstay Hunt, with "Places of Meeting" marked, shows the complete railway from Abbey to Llanymynech and Criggion, plus a branch from Kinnerley to Great Ness, crossing the Shrewsbury-Whittington road at Nesscliff. This branch was not built, but may have been included in the prospectus of one of the companies involved. The inclusion of proposed lines on maps at this time is by no means unknown.

Of the signalling equipment and methods used on the railway, little information has survived, but whilst clearly not elaborate they seem to have sufficed, as there is no record of any accident of consequence. Signal boxes were maintained at Shrewsbury (close to Abbey Foregate Junction) and at Kinnerley Junction, ground frames sufficing for the other stations provided with siding accommodation; of the disposition of signals only five are known: two near Llanymynech, one at Kinnerley, one at Hanwood Road and one at the Hereford line bridge which remained well into the dereliction period. The Criggion branch and the Llanyblodwel extension were unsignalled except at the junctions plus one signal at Llanyblodwel station; but there was one signal (with the reverse painted black) close to the London Road bridge on the Abbey Foregate spur. The junction at Llanymynech was controlled from the Cambrian signal box and indeed the lines in the deserted station were used by the larger Company as unofficial sidings after the closure of the "Potteries" system. Level crossings were provided with gates, as already mentioned, and in some cases cottages for gatekeepers were erected, as at Nesscliff, Shrawardine, Maesbrook and Criggion stations and the crossing near Melverley viaduct.

Such was the equipment, so far as is known, with which the ill-fated P.S. & N.W.R. started its struggle against adversity; on the whole, good solid construction which stood well the test of time when the Company had little or nothing to spend on even minor repairs. That the railway soon came on hard times was no reflection on the engineers who built it, but was attributable solely to economic factors, as will be seen in a later chapter.

Locomotives and rolling stock

Generally speaking, the aspect of railway operation upon which attention is focussed more than any other is the locomotives stock and it is therefore the more remarkable that information concerning the locomotives of the P.S. & N.W.R. is so fragmentary; nor does it seem likely that complete data will ever be forthcoming.

When the "North Wales Section" was ready for traffic, the Company had insufficient funds to purchase new locomotives and arrangements were accordingly made with the contractor, R. S. France, to supply the motive power. At the time of opening, the stock comprised ten engines, several of which had already been at work on the construction of the railway; but in alleviation of the acute financial crisis, five of these were put up for sale in December 1866, though, as the stock in 1867 is given as seven, it seems probable that only three were actually disposed of. Unfortunately, the Auctioneer's records of this transaction all went to salvage during the late war and the identity of the engines sold—and indeed some of the others—will, it is feared, have to remain in obscurity.

In 1865, Messrs. Manning Wardle & Co. supplied R. S. France with four locomotives, as under:—

Name	Type	Maker's No.	Cylinders	Wheels	Destination
Alyn	0-6-0ST	140	12″ x 17″ inside	3′- 0″	Llanymynech
Powis	0-6-0T	151	15″ x 22″ inside	3′-10″	,,
Sir Watkin	,,	167	,,	,,	Shrewsbury
Viscount	,,	168	,,	,,	,,

The first of these probably only worked on the construction of the P.S.N.W. for a short time, as it was transferred in 1865 to the Mawddwy Railway, for which R. S. France was also the contractor; it then became *Mawddwy*, later Cambrian No. 30 and later still G.W.R. 824, and was not cut up until October 1940, on the closure of the Van Railway, which it had worked for very many years. The other three engines did, however, work on the "Potteries" and two of them — *Powis* and *Bradford* (*Viscount* renamed) lasted till 1888. *Sir Watkin* was sold, probably in the period 1872-4 to

BEDFORD *in Walkden Yard, National Coal Board. August 22nd 1948. Formerly*
SIR WATKIN *of the P.S. & N.W.R.*   (*N. J. Allcock*)

Norton Cannock Colliery Co. Ltd., Staffordshire; resold by them about 1910 to Astley & Tyldesley Colliery Co. Ltd., Bedford Colliery, and named *Bedford*. She later worked at various collieries owned by Manchester Collieries Ltd., and finally gravitated to Walkden Yard, where she was scrapped in August 1950. Apart from the safety valves and the provision of a cab, it had not been much altered but some parts of the motion were stamped 151, indicating some rebuilding by the P.S. & N.W.R.

Two other locomotives of these early days later passed through the hands of Walker Bros. of Wigan, whose files provide information. One was *Nantmawr* a 0-4-0ST built by Henry Hughes of Loughborough in 1864; she went to Walker Bros. in 1873, and was resold by them two years later to I. W. Boulton, later becoming Girvan & Portpatrick Rly No. 4. The other was *Breidden* a four coupled well tank built by Hawthorns of Leith about 1865; she too was sent to Walker Bros. in 1873 and was rebuilt by them as a 0-4-0ST under their number 440. She was then sold to Messrs. Hall Bros. & Shaw, West Bank Alkali Works, Widnes.

The railway also obtained at least one 0-4-2 tender loco from the L. & N.W.R.; this was one of a batch of six built by Bury Curtis & Kennedy in 1847 for the Southern Division of that company and had 16″ x 24″ cylinders

17

5' 0" driving wheels and 3' 0" trailing wheels; it had a domeless boiler and on the "Potteries" retained its L.N.W.R. number, believed to be 1859. This engine, previously 1154 and before that 654, was sold to the P.S. & N.W.R. in May 1872, presumably replacing another engine. Its fate is unknown, but it had gone before 1875. Another engine which lasted until 1888 was the somewhat mysterious *Tanat*; it is known to have been a 0-4-2 T with 3' 6" driving wheels and 10" diam. cylinders, but the builder and other particulars have not been recorded.

The locomotive stock remained constant at seven between 1867 and 1872, but in the latter year a new engine was purchased, probably in replacement of one of the existing stock; this was *Hope*, a 2-4-0 T supplied by the Yorkshire Engine Co., Sheffield (maker's number, 185) and having 5' 0" driving wheels and cylinders 15" x 22". According to the Official returns, one engine disappeared in each of the years 1873/4/5 — but this does not agree exactly with the information regarding the two sold to Walker Bros. — the stock thus being reduced to the four (*Bradford, Powis, Tanat* and *Hope*) which remained to the sale of the P.S.N.W. rolling stock in August 1888. The final fate of these engines is not definitely known except for *Hope*, which went to the East & West Junction Rly, to be sold by them to B. P. Blockley, a dealer of Bloxwich, Staffordshire. He sold the locomotive in 1905 to the Cannock & Rugeley Colliery Co. Ltd., who gave it the number 8

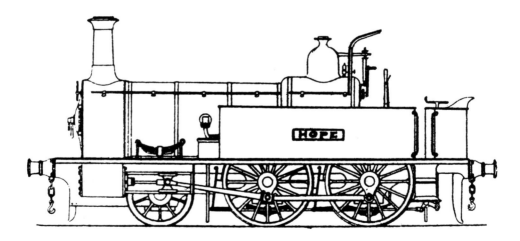

HOPE. *Yorkshire Engine No. 185 of 1872. New to the P.S. & N.W.R.*
(*Collection A. L. F. Fuller*)

and the name *Harrison* and rebuilt it as a 0-6-0T in 1916. This engine remained in service with the National Coal Board at Rawnsley until scrapped in 1955. The rather odd coincidence of two P.S. & N.W.R. locos going to places within a few miles of each other may be noted — *Sir Watkin* to Norton Cannock and *Hope* to Cannock & Rugeley. The date of sale of the loco to Norton Cannock is not known, only surmised — and she does carry some parts of *Powis*. If she was in fact *Powis*, not *Sir Watkin* she could have been sold about the same time as *Hope* by the same dealer Blockley, but this is purely guesswork. *Bradford* (Manning 168) has been recorded as later at the Blaenavon Iron & Steel Co., Pontypool.

Of the livery of the locomotives, no information has survived beyond the generalization that they were originally green and latterly black.

The coaching stock consisted of a primitive type of 4-wheeled 4-compartment coach with small windows and small comfort; 21 vehicles were in use in 1866, but five were disposed of at the end of the year and the remainder continued in use until 1877, when ten more disappeared. They were painted green at one time, but finished up a dull lead colour.

The goods rolling stock comprised 211 wagons at the opening, which was increased to 373 in the middle seventies and then dwindled to 265 at

HOPE at Rawnsley, in the ownership of the Cannock & Rugeley Collieries. By this time she had acquired a larger boiler and overall cab but was yet to be rebuilt as a six coupled engine. *(Collection A. L. F. Fuller)*

the closure; it consisted principally of six and eight-ton open wagons for carrying coal, lime, etc., a few timber wagons, cattle wagons, brakevans and a five-ton travelling crane. All, of course, have long since disappeared, though one lasted long enough to bear an S. & M. number.

The railway in operation, and its decline

At the time of opening, the advertised passenger service on the main line from Shrewsbury to Llanymynech consisted of five trains each way on weekdays, and two each way on Sundays, and friendly relations with the Cambrian Railways having been established from the outset, the trains were timed to connect with those of the other Company; on Saturdays, through carriages from Oswestry to Shrewsbury were run, with special cheap tickets. The first train, 6 a.m. from Shrewsbury, returning at 7.5 a.m. from Llanymynech, was a "Parliamentary" stopping at all stations and taking 55 mins. for the 18-mile journey; the rest of the trains were speedier, with a time allowance of between 40 and 50 minutes, accomplished by making some of the stops conditional; though the manner in which the initial timetable is besprinkled with these "conditional" stops, for which the Company had no experience to guide them, was arbitrary, to say the least. First, second and third class accommodation was provided on all these trains, in which respect the railway was ahead of most of its contemporaries. The Sunday trains, one "Government" and one ordinary, stopped at all stations and were scheduled to take a full hour for the journey, with one exception. For the winter timetable, the weekday service was reduced to four trains each way.

It will be noticed that speeds, though not of the express class, were not unduly slow, and making due allowance for stops, 30 m.p.h. must have been normal. The penny-a-mile rate was only available on the "Parliamentary" trains and the ordinary fares were on the high side, ranging from $1\frac{1}{2}$d. a mile for third class to $2\frac{3}{4}$d. a mile first class. Cheap tickets to Shrewsbury were issued on market days, however, and—by arrangement with the Cambrian-to Oswestry; further, day excursions to Aberystwyth and Borth and also week-end tickets from Shrewsbury to Aberystwyth, available for three days and by any train, were advertised.

The General Manager of the Company at the time of opening was Mr. J. Bucknall Cooper, who later became Manager of the Belfast Central Rly. In the middle seventies, he was succeeded on the P.S. & N.W.R. by Mr. A. Judd, who took over the unenviable task of trying to make the railway pay.

It was predicted that the railway would "prove acceptable to a rather extensive population", but one feels that perhaps "scattered" would have been a better choice of adjective. Even so, there seems some doubt as to

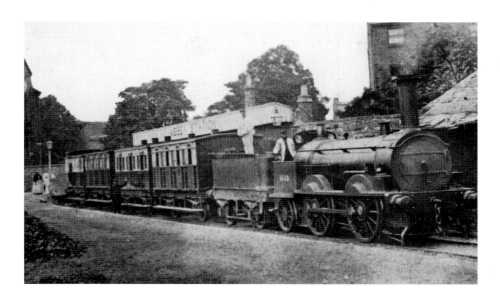

Bury 0-4-2 (ex L.N.W.R.) in Abbey station in the 1870's.   *(L. & G.R.P. 2706)*

whether the line was acceptable and before the year was out the Company was in trouble. One of their creditors, a principal of a London banking house, had demanded payment of his dues and, failing to obtain satisfaction, had put the bailiffs in possession of the line on December 3rd 1866; for the time being, trains continued to run with a travelling bailiff who remained in charge during the night. A meeting of Directors, Shareholders and Debenture Holders was called in London and as a result William Hall, the principal auctioneering firm in Shrewsbury, was instructed to sell by private treaty some of the Railway's equipment, comprising five locomotives, a number of wagons and other effects at the various stations on the main line. The announcement of this appeared on December 21st 1866 and thenceforward traffic ceased; and two years were to elapse before working was recommenced.

Things were still happening "behind the scenes" during this period; in the autumn of 1866, a scheme to amalgamate the Welsh railways (including the P.S. & N.W.) was put forward with the idea that, by co-operative action, the Welsh lines would be better able to fight the larger English companies in any proposals detrimental to Welsh interests. The various companies, being unable to agree amongst themselves, however, dropped out one by one and the project was abandoned.

The finances of the "Potteries" were straightened out during the ensuing couple of years and thoughts turned to the construction of the Market Drayton line, work on which had been commenced at Underdale (north-east of Shrewsbury) and elsewhere, but suspended in 1866; and in November 1867 application was made to Parliament to modify the existing Act, whereby the Abbey Foregate spur, instead of connecting with the Wellington line, would be built up to a higher level and carried over the latter to join the proposed Market Drayton line. It was considered that this connection between the "North Wales" and "Market Drayton" sections would be cheaper than that originally planned; and in fact some earthwork for this bridge were commenced but not completed. In the same Bill, further time in respect of the construction of the Llanyblodwel extension was also sought.

In December 1868, main line services were resumed on a reduced scale —three trains each way every week-day, with an extra on Saturdays, and one each way on Sundays; the Criggion branch was reopened on June 2nd 1871 and the Llanyblodwel extension in the following year; the 28 miles of single track then in operation remained thus until the closure. Branch passenger trains to and from Criggion met all main line trains at Kinnerley and the latter were retimed to and from Llanyblodwel terminus.

In 1866 the railway had worked at a slight loss (£99) and it was not until 1871 that a profit was shown, whilst 1872—perhaps as a result of the economies of single line working—proved the most successful year, with a profit of £1,019. The smiles of Fate were, alas ! but fleeting ; in 1873 the line just paid its way but in each of the three succeeding years a loss of over £2,000 was recorded. Further steps to curtail expenditure were put in hand, including the reduction of station staffs which, coupled with the single line working, tended to reduce the speed of trains and generally impair the service to the public, thereby effectively countering the attempts made at the same time to stimulate traffic. The latter included the abolition of second class travel; the reduction of third and first class fares to 1d. and 2d. a mile respectively and the issue of first class return tickets at single fare and a third; and the running of special cheap excursions to Llanymynech and Criggion.

As a result of these measures, the annual deficit was halved, and the passenger traffic improved, but no real recovery was made; the Company steadily descended into ever worse financial straits, by 1877 causing serious alarm and culminating in the appointment of an Official Receiver. The railway became a butt for public humour and many amusing tales from these difficult days are recalled by older people; when the financial trouble became acute, angry creditors put the bailiffs in and on one occasion the latter refused to allow a train to start from Shrewsbury, but eventually relented on condition that they were allowed to travel in it — only to be quietly "dropped" at the first available station under the cloak of a shunting operation.*

The reasons for the decline in fortune of the railway are not far to seek, and indeed it is rather surprising to learn that the line ever paid at all. Other than Shrewsbury, none of the places served were of any commercial importance, even Llanymynech boasting less than a thousand souls and the rest merely tiny villages or hamlets; three stations—Red Hill, Hanwood Road and Ford—lay on important roads radiating from Shrewsbury, but had nothing else to commend them.

Scenically the railway had a good deal to recommend it, for the line passed through some delightful green countryside and was a convenient route to a number of Shropshire hills popular with walkers and pleasure parties: Llanymynech Hill, the Breiddens at Criggion and Nesscliff Hill, where there is still to be seen the cave once the stronghold of Humphrey Kynaston, the legendary 16th century robber chieftain. At week-ends and holiday periods the railway enjoyed a considerable traffic from trippers and also from fishermen who had their favourite haunts on the Severn, as at Shrawardine, and the Vyrnwy; but these were not enough to make the "Potteries" a paying concern and the railway could only be regarded as a purely cross-country line whose chief function was the transport of stone from the quarries at the termini. Upon this mineral traffic the fate of the railway hung and any reduction below a certain minimum simply ruled the line out as an economical proposition. The closing of a few quarries and the failure of others to take their place (it had, for example, been anticipated that long-disused quarries at Belan on the Criggion branch would be re-opened) was the direct cause of the railway's misfortune.

Once started on the downward path, the decay of the railway was increasingly rapid. In a subsequent enquiry it was disclosed that from the inception of the scheme in 1862 up to the end of 1877 no less than eleven Acts of Parliament had been granted in respect of the undertaking and a capital expenditure of almost £1,400,000 incurred, i.e. about £50,000 per mile, a disproportionately high figure; after 1877, no one seemed to care much what happened. The staff was reduced to the absolute minimum, tickets for instance being issued by the guard; the track was overgrown with weeds and the hedges unkempt; and the rolling stock and the equipment generally presenting a very sad spectacle of disrepair. The Melverley viaduct was already showing signs of feebleness, and the Company took the precau-

---

* This story is believed to be quite authentic, but another almost precisely the same is sometimes told of the Bishop's Castle Railway, however, and too much credence should not be placed in them, as confusion between these two lines is by no means uncommon. The writer had related to him a vivid description of a "double-ended railcar"—clearly one of the S. & M. contrivances—operating on the B.C.R.: but there is no evidence that any of these cars ever left their home metals, and it seems that the teller had confused the two railways, which did, after all, have a good deal in common from the public viewpoint.

tion of running branch trains only on Wednesdays and Saturdays (four each way) in connection with the main line service of four trains; and one on Sundays connecting with the up and down main line train.

By the beginning of 1880 things had come to such a pass that the Board of Trade Inspector (Colonel Rich) imposed a speed limit of 25 m.p.h. pending track renewals. The end, it was apparent, was at hand; and yet, at this late hour when all seemed lost, rescue was very nearly effected from a totally unexpected quarter. The Great Northern Railway had for some years been pursuing a vigorous policy of expansion and had built a line to Burton and another through Derby making connection with the North Staffordshire Railway; over the latter system they obtained running powers to Uttoxeter, thence to Stafford in 1879 by arrangement with the struggling Stafford & Uttoxeter Railway (which they absorbed in 1881) and were now seeking to obtain running powers to Shrewsbury and also, via Abbey Foregate Junction and the "Potteries" to gain access to the Cambrian Railways and the Welsh coast. In this scheme to bring Wales within easier reach of the industrial East Midlands they were actively aided and abetted by the Midland Railway, to the manifest annoyance of the Great Western and London & North Western Companies.

On the other hand, the people of Shrewsbury, seeing the promise of fulfilment of the original intentions of the P.S. & N.W.R., were emphatically in favour and following a meeting of the Town Council in May a resolution to that effect was sent to Parliament. If the proposals were carried out, Alderman Southam said, " It would probably allow that line to be more efficiently worked and perhaps repaired a little oftener "; and in any case " any change would necessarily be for the better." On the 11th of June 1880 the local press announced the substance of the proposals and instanced the saving of time that would accrue from the through running envisaged, and at long last it seemed that there was some hope for the poor old "Potteries", when out of nowhere came an order from the Board of Trade demanding that—no track repairs having been carried out—the line must be closed to traffic in the interests of public safety. Accordingly, without previous notice, trains on the P.S. & N.W.R. ceased as from June 22nd, 1880, under the customary euphemism of "service suspended"; and everything was literally left to rot where it stood.

It is permissible to speculate on the possibility of a successful outcome of the G.N.R.-M.R. compact, but there are grave doubts whether the actual closing of the "Potteries" had any material adverse effect on the scheme; the line was in such a bad state in all respects that from the angle of reconstruction expense it could hardly have mattered to the Great Northern whether they obtained the concern alive or dead. It seems probable that the

Former P.S. & N.W. Guard Reeves (in shirt sleeves) with his patrol trolley on line west of Hanwood Road. The wooden central portion of bridge was rebuilt with brick by Shropshire Railways, but parapet not completed.

(F. E. Fox-Davies, L. & G.R.P. 2817)

Llanymynech station, with water tower and loco shed in background. November 1902.
(F. E. Fox-Davies, L. & G.R.P. 2813)

G.N.R.-M.R. proposals were unworkable on other grounds; and so the "Potteries", surely one of the most ill-starred of railways, was left to sleep in peace.

The Shropshire Railways
Period 1880 - 1907

Formation of the new Company

FOR eight years the solitude of the "North Wales Section" was undisturbed save for the leisurely ministrations of the two or three men appointed by the Official Receiver' ostensibly to maintain the fences in order and generally keep an eye on the property, for which purpose they were provided with a hand-operated trolley. Amongst them was one Richard Reeves, at one time a guard on the "Potts", and to one who knew the line in its palmier days his must have proved a meloncholy task; the track was overgrown long before the closure and the process of decay, with a running start, continued apace and soon took heavy toll of the rolling stock and equipment. The four locomotives *Bradford*, *Hope* (sic), *Tanat* and *Powis* suffered the least, as they were all housed in the Shrewsbury shed and were maintained in reasonably good condition by the "staff"; but the job of looking after the rest of the rolling stock, distributed all over the

Kinnerley station, looking east in November 1902. Note the haystack on line under bridge and the hand trolley.　　　　　　　　　　　　　　　(*F. E. Fox-Davies*)

27

system, was far beyond their powers and most of it was left simply to fall to pieces. The coaching stock (such as it was) stood in a siding at Abbey station in a very dilapidated condition which the passage of time only intensified; before long, all the windows were smashed, doors wrenched off and the dusty interiors littered with broken fittings.

The goods wagons were left where the Closing Order found them, which was at practically every station on the system, but the biggest concentration was at Abbey Foregate station and on the spur to the Wellington line.

The track became increasingly verdant and the hedges grew rank, the flooring of bridges rotted and fell away, and buildings—the wooden ones particularly—soon showed signs of dilapidation; but the permanent way itself stood up to its test fairly well. For eight years the railway mouldered and apart from the results of decay the only change in that period was the removal of the junction of the Abbey Foregate spur with the Wellington line, from which point the rails were lifted in the early eighties and a fence placed across the track.

The "Potts" however was not dead, but merely sleeping; and following an agreement made on March 14th 1888 plans for its resuscitation were submitted to Parliament under the style of the "Shropshire Railways Bill", which also aimed to build a line from Shrewsbury to Hodnet and thence, via running powers over the Great Western to Market Drayton and over the North Staffordshire beyond, to reach Stoke. As usual, Salopian support was enthusiastic, tradesmen with premises near the old line having suffered to some extent by its demise and all being in favour of a more direct route to the Stoke area, whither the most redoubtable advocate of the railway, Alderman Southam, had journeyed to ascertain local feeling in the matter and had reported his favourable findings to the Town Council in March. A resolution supporting the Bill was unanimously passed and at a public meeting held on June 5th 1888 the Mayor announced that £3,500 had already been subscribed to meet preliminary expenses, and Mr. Alex Young, the Liquidator of the P.S. & N.W.R., that the shareholders of the Company had agreed on March 14th 1888 to accept £350,000 shares in the Shropshire Railways in the discharge of their dues on a percentage basis. The Great Western ridiculed the whole scheme and, it must be conceded, not without some justification; it was indeed agreed that the Market Drayton extension was the cornerstone to the success of the venture. The Engineer, John Russell, C.E., estimated that £100,000 would be required, of which £41,000 would be spent on reconditioning the P.S. & N.W.R. — £27,000 for the main line, £3,000 for the Llanyblodwel extension and £11,000 for the Breidden branch. In due course, the Bill passed the House of Lords and on Tuesday, July 24th 1888 came before a Select Committee of the Commons, under the Chairmanship of Sir Jno. Kennaway; the Committee after a short session gave the Bill their blessing, and the Company received their Act of Incorporation August 7th 1888, with the qualification that in deference to the Great Western, Hodnet was to become a second Redhill, running powers to Market Drayton having been turned down. The capital of the S.R. Co. for the "Shrewsbury Separate Undertaking" consisted of 350,000 shares and 150,000 debentures in round figures.

The fate of the line thus decided, the P.S. & N.W.R. made steps to dispose of the locomotives and rolling stock, which the new Company did not wish to take over, and the Liquidator arranged that the firm of Arthur T. Crow of Sunderland should sell the movable effects by Public Auction on Friday, August 24th 1888. The sale, which attracted a considerable number of buyers, was held by the upper railway bridge in Abbey Foregate, as the bulk of the stock was accumulated at this end of the line. Worthy of note were two "conditions of sale" which directed that each purchaser pay the auctioneer 5 % of the lot money and remove his property at his own expense within six days; the latter proviso was indeed severe, as there were more than 150 vehicles to be moved (over track decayed and overgrown and points wellnigh solid with rust) and the only outlet was via the yard of the Midland Wagon Co. and on to the G.W.-L.N.W. line at Coleham.

The locomotives were first disposed of (the sale being conducted at

Nesscliff station (left) with Cross Keys Cottage (right) , looking west in June 1903. The cattle pens were the only ones remaining on the railway. The rails can just be seen in the foreground. (*F. E. Fox-Davies, L. & G.R.P. 3902*)

Shrawardine viaduct. Probably pre-1902.   (F. E. Fox-Davies, L. & G.R.P. 2829)

the engine shed) and, having been under cover and given some measure of attention, realized quite good prices, as follows:—*Tanat*, £85; *Hope*, £200; *Powis*, £210; and *Bradford*, £300, through the last-named was stated to be "not complete".

There is no record of the sale of any of the passenger stock, and the inference is that the coaches were simply worth nothing more than breaking up for firewood; but the goods wagons fetched prices from £3-10-0 (for 6-ton open wagons) upwards. There were two goods brakevans, two cattle trucks, ten timber wagons, and a 5-ton travelling crane, all the rest being open wagons for carrying coal or lime - mostly 6-tonners, but including a few carrying 8-tons. When all the stock at the Shrewsbury end had been sold, the party moved up the line to the various stations where wagons had been stranded at the closure; and by the end of the day, all the rolling stock of the "Potteries" had been sold save for one 6-ton wagon at Kinnerley (one of half a dozen here) which remained until the coming of the S. & M.R.

The Reconstruction

The way was now clear—or so it seemed—for the Shropshire Railways to go ahead with its plans; but all sorts of difficulties, chiefly financial, kept cropping up. The Company obtained their Order on July 17th,

Melverley viaduct before collapse. Pre-1902.   (*Collection N. J. Allcock*)

1889, amended in the following month, and it then only remained for them to settle with the shareholders and creditors of the "Potteries"; this process took longer than anticipated, however, especially as one gentleman, less altruistic than the Earls of Powis and Bradford and other landowners, insisted—quite unreasonably, it is said—on being paid in cash; but at the General Meeting held at Westminster in June 1890 the Chairman, Sir Richard Green Price, announced that these difficulties were being overcome and finally, after two years of the most heartbreaking wrangling, the promoters obtained possession of the sorry remains of the P.S. & N.W.R. on September 19th, 1890.

The Contractor, Messrs. Charles Chambers of Westminster, was instructed to proceed forthwith and had undertaken to complete the work of rehabilitation within twelve months; this involved the relaying of the main line to Llanymynech, the reopening of the Abbey Foregate conne tion (to which the Joint Companies had signified assent), the raising of Abbey station above flood level and the construction of an approach to it better than the existing sharp descent. It will be observed that the Company's immediate plans did not include the reopening of the Breiddon branch, they having decided the Market Drayton extension to be more important and to proceed with that as soon as the "Potteries" main line was reopened.

For nearly twelve months the task of reconstruction was actively pushed

31

forward; almost the whole of the main line was resleepered and refenced and work on bridges, which included the replacement of timber structures by new ones of brick and concrete or girder type, was well advanced; a start was also made in raising Shrewsbury station, a quantity of earth being deposited on the site and a new bridge built over the Rea at a higher level. The work still awaiting completion was chiefly in connection with stations-yards, sidings, passing loops, level crossings, etc.—when, towards the middle of 1891, financial difficulties reasserted themselves, this time in a more serious form.

The firm of Whadcoat Brothers Ltd. had entered into an agreement to finance the Contractor, whilst the brothers themselves, J. H. and W. E. Whadcoat, were at the same time debenture holders in the Shropshire Railways; the delay in obtaining possession of the property meant that debenture holders could only be paid out of the funds, which the firm of Whadcoat refused to do, and the Directors paid them out of their own pockets. Moreover, Whadcoat Bros. Ltd. ceased payments to the Contractor, who perforce withdrew his men on July 15th, 1891. The Shropshire Railways sued the solicitors for the moneys witheld, and the latter made an application in Chancery for the appointment of an Official Receiver. The proceedings seem indeed to have been far from regular and there was evidence too that the Whadcoats were trying to obtain control of the Board by cornering shares; but in August, Justice Chitty ruled that an Official Receiver must be appointed and on November 11th, 1891, Robert Rabbidge took charge on behalf of the debenture holders. The final outcome of the legal quibbling, however, was never decided, as the Company had no money left with which to fight. A Bill providing an extension of time for the completion of the Market Drayton extension had received the Royal assent on July 21st, 1891; and another Bill, promoted at the same time for the construction of a line along the Tanat Valley, passed the House of Lords and the Committee stage of the Commons only to be withdrawn for lack of the necessary £5,000 deposit. Weighed down with such a heavy financial burden, the Shropshire Railways gave up the struggle, the works were abandoned, and once more the ill-starred line settled down for a long rest.

The period of dereliction

As recorded, 1891 saw the end of the Shropshire Railways Company so far as outward activity was concerned and apart from the insidious toll of the elements the railway remained for nearly twenty years much as the Contractor had left it. It will be recalled that the latter had concentrated on the permanent way, the 54% of the contractual work done including very little attention to buildings, which had already been rotting for more than a decade; and by the turn of the century the whole railway provided a

Crossgates station, looking west. June 29th 1903.   (L. & G.R.P. 3893)

very sorry spectacle of desolation. Fortunately for posterity, a vivid description of the line in its declining years has been preserved in the classic article from the pen of the well-known railway historian T. R. Perkins*, who, in 1902 traversed on foot the whole of the system with the exception of the impassable section from Kinnerley to Melverley viaduct. As one who knew something of the line in earlier days, he found much to ponder over and certainly the forsaken remains would have stirred the imagination of one less receptive.

Abbey station had been converted into a stable, the space between the platforms filled in and the whole used as a coal dealer's yard; part of the parapet of the new Rea bridge was lying in the stream together with the girders of the old; most of the buildings had been demolished, including the loco shed and signal cabin at the junction, whilst the angle between the lines here was used as a tip. The permanent way on the lower part of the incline from the station was buried and on the upper half torn from the sleepers, but beyond that was complete throughout though overgrown; that on the Criggion branch (which had not been touched since 1880) was completely buried in a tangle of bushes and undergrowth. Some of the bridges had received attention from the Shropshire Railways Co., and though the wooden floorings rotted away the girders remained intact with

---

* "A derelict British Railway" by T. R. Perkins; the *Railway Magazine*, 1903, pp.400, 441.

Wern Junction in 1904, looking towards Llanyblodwel. The Cambrian Railways line to Llanfyllin is on left: their original line crossed on the dismantled bridge beyond the gate.
(L. & G.R.P. 3917)

Llanyblodwel station, P.S. & N.W.R., looking towards Nantmawr; rebuilt as Blodwell Junction by the Tanat Valley Light Railway. The building on the left is a coal office.
(L. & G.R.P. 3916)

the exception of those crossing the Hereford line, which were removed; the Shrawardine viaduct was well preserved but that at Melverley, already tottering before the P.S. & N.W.R. closed, simply fell to pieces and was carried away on the current, though four of the original piles survived to the 1950's.

The station buildings suffered most of all, those at Hanwood Road, Red Hill, Shrawardine and Maesbrook disappearing entirely and the rest assuming a very ruinous condition, with windows broken and old notices peeling off the mildewed walls; each a home for birds and mice, who made freely with the waybills and rotting paper to build their nests. At Llany-mynech, the booking office door had come from its hinges and was propped against the wall, exposing the interior with the books and fittings littering the floor, whilst a similar atmosphere of decay pervaded the adjacent yard; the old locomotive shed and carriage shops were st ll standing, together with the water tower and the turntable, now choked with bushes and trees fifteen feet high. Similar scenes of desolation were to be seen at every station; only three signals were still standing, and another lay on its side close to the ruinous and inclining box at Kinnerley junction, together with the water tank.

The old "Potteries" servant, Richard Reeves, was retained by the Official Receiver to · maintain the fences, but his position must have been something of a sinecure; most of the route was bounded by hedges—by now almost qualifying for the title of spinneys—and the level crossing gates were replaced by hurdles of brushwood as they fell to pieces. He still used the hand-operated trolley to propel himself along the line, a job which could be relied upon to keep anyone warm on the coldest of days, but the passage over the metals was by no means easy and the way lonely; one or two of the crossing cottages were still occupied, others deserted, and here and there the railway would be used by a local farmer to place a haystack. The Criggion branch was naturally in an even more dilapidated state than the main line, from the bay at Kinnerley choked with bushes to the tiny passenger ter-minus at Criggion, standing amid a dense thicket in the middle of a pasture; the line continued to the quarries beyond the station, a distance of about a mile.

In contrast to this mouldering immobility, the Llanyblodwel extension saw a number of changes over this period. The Cambrian Railways' Llany-fyllin branch (opened July 17th 1863) left Llanymynech to the north of the station, facing Oswestry, and crossed the P.S. & N.W. line before running up the Afon Cain valley to Llanyfyllin. By an agreement with the Receiver dated January 20th 1881, the Cambrian Railways undertook to maintain and operate the P.S. & N.W. line between Llanymynech and Nantmawr quar-ries (the terminus of the Llanyblodwel extension) and this was reopened for traffic on January 1st 1886, following which they abandoned the first mile or so of their Llanyfyllin branch in favour of the "Potteries" line from Llanymynech by putting in a connection (opened January 27th 1896) between the two at Wern. The length of "Potteries" line taken over was about 55 chains. By this means the journey was shortened and the need for reversal and some steep grades and sharp curves eliminated. The lease for working the nearly four mile Nantmawr line was renewed by agreement with the Shropshire Railways dated July 1st 1900, the annual rent being £555. This sum was increased to £705 by a Supplementary Agreement dated December 30th 1915 and on September 1st 1939 increased again to £886: all of which was paid to 5% Mortgage Debenture holders.

One other development in this area remains to be mentioned—the Tanat Valley Light Railway, opened in 1904. This railway has no direct bearing on the history of the Shropshire Railways, being promoted by a group of the local gentry to serve the slate quarries at Llanygynog and the agricultural district bordering the Tanat, but is of interest as lying on part of the original route selected for the stillborn West Midland, Shrewsbury and Coast of Wales Railway (see Page 8). The line was a prolongation of the Cambrian Railway's Porthywaen mineral branch and utilized a few chains of the old "Potteries" line in the vicinity of Llanyblodwel station, which was renamed Blodwell Junction; the P.S. & N.W.R. lines diverged southwards to Llanymynech and northwards to Nantmawr respectively west and east of this station. After the opening of the new line, a morning train was run between Llanymynech and Blodwell Junction, but this service soon ceased. The Tanat Valley Railway was worked by the Cambrian by agreement, finally being absorbed by them in 1921; later passing into Great Western hands, it became of note by reason of the strange miscellaneity of the motive power supplied for its operation—ex Cambrian, Alexandra Dock, Liskeard & Looe and Whitland & Cardigan engines all being used.

From this time onwards the portion of the "Potteries" west of Llanymynech ceases to be relevant to this history and latterly there were only two portions of it in use — the first mile of the Llanfyllin branch and the last mile from Blodwell Junction to Nantmawr, then a branch of the Tanat Valley line; the intervening track was lifted, and today, of course, all has gone. The rest of the "Shropshire Railways" meanwhile lay moribund, awaiting the next development.

The Shropshire & Montgomeryshire Railway

Period 1907 - 1941

Promotion of the Light Railway

THE miserable end of the "Potteries" and the ineptitude of its successor would, one might have thought, have deterred any further effort to run trains between Shrewsbury, Llanymynech and Criggion. But no; from time to time in the early 1900's resolutions were passed by various local councils pressing for the reopening of the line and these voices eventually found a sympathetic ear in that most ardent believer in the light railway, Mr. H. F. Stephens, crystallizing in the promotion of the Shropshire & Montgomeryshire Light Railway. This scheme was from the outset much less ambitious than any of its predecessors, the sole aim being to revive and work the existing Shropshire Railways system, Colonel Stephens (to use the title by which he is best known) considering that the reduced expenditure concomitant with light railway operation would enable this to be done on an economic basis; that his optimism was not altogether shared by the G.W.R. and L.N.W.R. seems clear, however, from the fact that these Companies refrained from raising any objections this time — presumably they held the view that it was no business of theirs to interfere if yet a third Company chose to throw good money after bad. Local support for the new venture was enthusiastic and in the application to the Light Railway Commissioners dated May 30th 1907, the following authorities asked for permission to advance (as part of the share capital and/or as a loan secured by the issue of Shropshire Railways prior charge debenture stock) the sums stated; Shropshire County Council, £2,000; Montgomeryshire County Council, £1,500; the Corporation of Shrewsbury, £500; Oswestry Rural District Council, £250; Forden R.D.C, £500; Atcham R.D.C., £1,000. A local Enquiry satisfying the Commissioners as to the desirability of granting the Order, and the draft approved by the Official Receiver of the Shropshire Railways on January 17th, 1908, the Shropshire & Montgomeryshire Light Railway Order was issued by the Board of Trade on February 11th 1909, under the Light Railways Act of 1896.

The new company acquired the effects of the Shropshire Railways on relatively easy terms, and though nominally the latter company had considerable powers in deciding policy, no objection was raised to the course of action proposed by Colonel Stephens. The authorized capital was £2,000 in £10 shares and the Shropshire Railways were empowered to issue 4½% prior charge debenture stock to nominees of the S. & M.R. up to £32,000 in respect of the main line and £8,000 for the Criggion branch, the holders of such stock to be entitled to vote at S. & M.R. meetings. The S. & M.R. could use but not purchase lands owned by the Shropshire Railways; nor could they touch the existing revenue from the latter or the money in the Market Drayton Deposit Fund. The loans of the various local councils were repayable in 40 years. There were three Directors; the Right Honourable the Earl of Bradford, William Rigby and Holman Fred Stephens, though the Shropshire Railways had power to nominate two more directors if they thought fit.

The railways which the new company were authorized to operate were the main line; the Criggion branch, subject to Shropshire Railways approval; the Abbey Foregate spur; and a connecting line with the Shrewsbury-Hereford line close to the Midland Wagon Co's works. The last-mentioned had been authorized under the Shropshire Railways Act of 1888 but had not been built; neither was it nor the spur under the S. & M. regime. In accordance with the provisions of the Light Railways Act, simplified methods of working were permitted, subject to certain restrictions. Signals needed only to be placed at stations, for the protection of passing loops; no turntables were necessary; and gates at level crossings could be

The contractors locomotive used in the refurbishing of the line in 1910, standing on a L.&N.W.R. well wagon in the Shrewsbury Exchange sidings.   *(H. Miller)*

dispensed with provided that cattle guards and the usual warning notices were erected. A speed limit of 25 m.p.h. was to apply (reduced to 10 m.p.h. on curves of radius less than 9 chains and within 300 yards of an unprotected level crossing) and a maximum load per axle according to the weight of rail employed. A simple fare scale was envisaged, i.e. 1d. a mile 3rd, 2d. per mile 2nd and 3d. a mile 1st class; and the company could, and did, charge a supplementary fare equivalent to two miles for conveying a passenger across either of the Severn viaducts. The work of reconstruction was to be commenced within 12, and completed within 18, months of the granting of the Order unless extended time were approved by the Board of Trade; and if it was decided to reopen the Criggion branch, the permission of the Shropshire Railways was to be obtained within two years and the work completed within a further 18 months.

Construction and equipment

A month after the confirmation of the Light Railway Order, a private meeting of potential supporters of the scheme, with the Earl of Powis presiding, took place on March 19th 1909, with the result that a small committee was formed to visit the East Kent Light Railways (then under construction) and the Kent & East Sussex Railway, to both of which Colonel Stephens was Engineer; their favourable findings did much to influence local support for the railway. Nevertheless, some delay did ensue and it was not until well into 1910 that the work of reconstruction was commenced from the Llanymynech end, where the double junction with the Cambrian was the only physical connection of the Shropshire Railways with the outside railway world.

Clearing a path through the jungle was the first job; after the tangle of weeds, brambles and overgrown bushes had been cut back and burnt at intervals, the contractors, using a six-coupled saddle tank engine (of Manning Wardle design) were able to get down to the principal task, the relaying of the track. The original 1865 24' 0" rails and chairs were found to be in good condition and in the main they were torn bodily from the rotting sleepers and reutilized; but a few chairs were replaced by second hand ones from the S.E. & C.R., S.E.R. and C.K. & P.R., dating from the early 1900's; the sleepers laid in 1888 were uncreosoted and in so bad a condition as to require complete replacement—indeed, the original P.S. & N.W.R. sleepers on the Criggion branch and a few other places were in better shape. Apart from modifications in some of the station layouts, the only new track laid was a connection, with sidings, to the Welshpool line near Meole Brace, about two miles from Shrewsbury. The sole engineering work of any magnitude was the new viaduct at Melverley, a girder structure laid on wooden piers, placed a little to the east of the old viaduct, long since

39

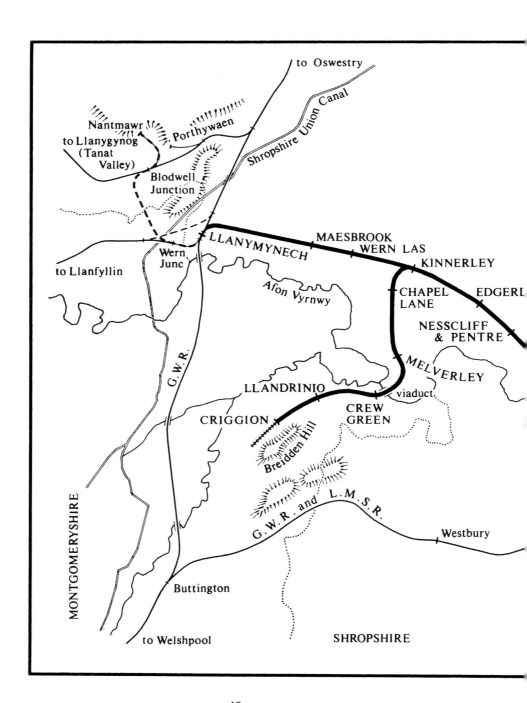

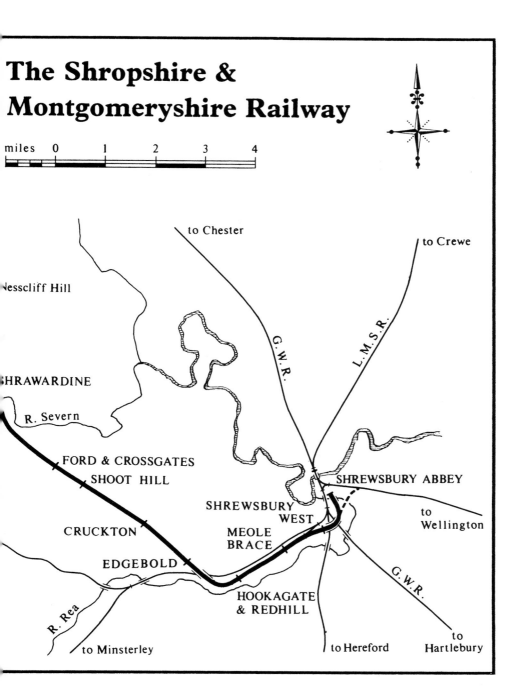

The Shropshire & Montgomeryshire Railway

miles 0   1   2   3   4

to Chester

to Crewe

Nesscliff Hill

G.W.R.

L.M.S.R.

HRAWARDINE

R. Severn

FORD & CROSSGATES

SHOOT HILL

SHREWSBURY ABBEY

SHREWSBURY WEST

CRUCKTON

MEOLE BRACE

to Wellington

EDGEBOLD

G.W.R.

R. Rea

HOOKAGATE & REDHILL

to Minsterley

to Hereford

to Hartlebury

41

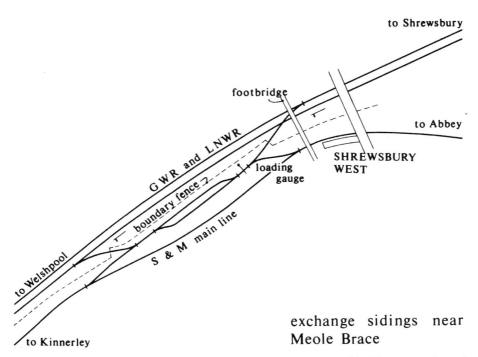

to Shrewsbury

footbridge

to Abbey

GWR and LNWR

boundary fence

loading gauge

SHREWSBURY WEST

S & M main line

to Welshpool

to Kinnerley

exchange sidings near Meole Brace

carried out to sea; otherwise, no extensive rebuilding of bridges was found to be necessary, but in a few instances girders from the lifted track were taken up and used in strengthening the existing line.

A similar economy of materials characterized the rehabilitation of the stations; brick buildings, where they existed, were cleaned up and used again and even some of the wooden ones were refurbished, as at Llanymynech, Ford & Crossgates and Criggion; new wooden buildings, comprising a stationmaster's (or, as it sometimes happened, stationmistress's) room, booking office and waiting room, were erected at Edgebold (the former Hanwood Road), Shrawardine and Maesbrook. Red Hill was replaced by a new station named Hookagate & Redhill about 100 yards nearer Shrewsbury, and of similar type to the others. Nesscliffe was renamd Nesscliff & Pentre, and Crossgates (originally Ford) became Ford & Crossgates. A number of new stations and halts, consisting mostly of simple brick platforms, were built at Shrewsbury West (1 mile from Shrewsbury, close to the Welshpool line); Meole Brace ($1\frac{3}{4}$ miles; with wooden booking office); Cruckton ($5\frac{3}{4}$ miles; adjoining the Montgomery Road, and with shelter on the platform); Shoot Hill ($6\frac{3}{4}$ miles); Edgerley ($12\frac{1}{2}$ miles; a platform of sleepers by a dismal pine wood); Wern Las (15 miles); and one, Chapel Lane at a level crossing on the Criggion branch $\frac{3}{4}$ miles from Kinnerley. It should perhaps be mentioned that the changes listed in this paragraph did not all take place from the opening of the line, but were brought about gradually; Cruckton, for example, though shown in the first timetables, did not come into use till some time afterwards.

Shrewsbury station was not materially altered, the erstwhile stable being converted back to its original function, comprising traffic office, booking office, waiting room and lavatory, with the new platform beyond the buildings on the higher level chosen by the Shropshire Railways; a bay platform and a goods yard were laid alongside. Passing places were provided only at Ford & Crossgates (with three sidings) and Kinnerley, where a goods yard was built and which also retained its bay platform; Llanymynech preserved its double platform and two sidings besides. Shrawardine had a loop siding and one other; Nesscliff & Pentre a loop siding and two more; the other stations, excluding the three "halts" at Shrewsbury West, Shoot Hill and Edgerley, had one siding each. Criggion had a run-round loop, the line continuing beyond to the quarries of the Pyx Granite Company, through a gated level crossing.

The original locomotive shed at Shrewsbury had been pulled down by the Shropshire Railways Company in their attempt to improve the approach to the station—the site later being covered by part of the Midland Wagon Co's works—and though the remains of the other shed still stood at

Hookagate station, looking west, and showing conditional stop signal. May 20th 1929.
(R. K. Cope)

43

Llanymynech, the new company decided to have one central loco shed and repair shop at Kinnerley Junction instead, and the Llanymynech building was ultimately dismantled and the turntable pit filled in. The new shed was a two-road structure capable of housing six tank engines, and was laid out with inspection pits and a machine shop attached, adjoining the Criggion branch a short distance from the junction; alongside were situated a small yard and two other buildings for the repairing and painting of locomotives and rolling stock. Water for locos was obtained from an iron tank on tall brick supports, the supply being obtained from a deep well by means of a windpump, the vanes of which formed a familiar Kinnerley landmark for many years; in addition, water tanks were maintained at Shrewsbury, Shoot Hill and Llanymynech. The original tank at the lastnamed was burnt down in 1904, probably as a result of sparks blown from a Cambrian engine in the adjoining station.

As laid down in the S. & M. Act, semaphore signalling was confined to stations with passing places and there were accordingly four ground frames, supplied by Messrs. Tyers-Shrewsbury (8 levers, including 2 spare): Ford (7 levers, including 1 spare): Kinnerley (13 levers, including 1 spare): and Llanymynech (5 levers, including 1 spare). There were also signals facing each way at Meole Brace, where tickets were examined. All stations were connected by telegraph and the sections Shrewsbury-Ford and Kinnerley-Llanymynech were operated on the Staff & Ticket system, the

The level crossing north of Melverley viaduct. The original P.S. & N.W. gate-keeper's cottage on right; cattle guards by S. & M. May 14th 1948. (E. S. Tonks)

intervening section by Electric tablet and the Criggion branch on the "one engine in steam" principle. Points at minor stations were opened by means of Annett's Key attached to the train staff. These were only "conditional" stops for passenger trains, and for the guidance of drivers a large diamond-shaped board, painted red with a broad white band, was placed on the platform of each. In the "off" position, the board was parallel to the track, but could be rotated to the position at right angles if a train was required to stop; at night, red and green lamps showed appropriately.

The Shropshire & Montgomeryshire Railway at work

The formal opening of the railway took place on Thursday, April 13th, 1911, and at the invitation of the Managing Director, Mr. H. F. Stephens, some 200 guests assembled at Abbey Foregate station to participate in a ceremonial run over the line; amongst these were Lord Forester, the Deputy Mayor of Shrewsbury (Alderman Ben Blower, who had done so much to raise money for, and influence public opinion of, the railway), representatives of the various subscribing local councils, the directors of the Company and various other friends of the railway, including—a happy thought—the railway historian who had evinced such keen interest in the line at all times, Mr. T. R. Perkins.

The train consisted of the six-coupled locomotive *Hesperus*, its smart

The second PYRAMUS *with a passenger train consisting of ex L.&S.W.R. coaching stock.*
(Real Photographs Co. Ltd.)

45

olive green livery sparkling, four bogie coaches, two four-wheeled brakevans and (for the outward journey) two Cambrian saloons, one of which had been used by the late King Edward VII on the trip over the Birmingham Waterworks Railway, Rhayader, in 1904. The proceedings were opened by the Mayor of Shrewsbury, Major Wingfield, in the form of a short speech delivered from atop the leading carriage, the while he held up the Loving Cup of the Borough of Shrewsbury. " We are assembled here," the Mayor said, " to open the Shropshire & Montgomeryshire Light Railway, a railway which I trust and think will be of great benefit to this borough and the country districts which it serves between Shrewsbury and Llanymynech . . . It is a kind of mutual benefit railway which I have the honour of opening to-day. I hear the engine blowing off steam, so for fear it should burst I curtail my remarks and drink of the Loving Cup ' Success to the Shropshire and Montgomeryshire Light Railway '." The Mayor then took his seat on the train and the latter moved out of the station to the accompaniment of fog signals, flag waving and cheers from the hundreds of sightseers that had gathered at the terminus and along the line all the way to Shrewsbury West station; the 1 in 47 gradient out of Abbey station would without doubt provide a sufficiently impressive sight, with *Hesperus* and eight coaches. At Ford the party disembarked to be photographed by Councillor R. L. Bartlett and then proceeded to Kinnerley " running with a smoothness which would compare not unfavourably with some of the greater railways of

Abbey station, probably in 1920's. Note Sentinel steam lorry in yard.
(F. Moore. 1664)

the country*"; it is to be feared that in later years the railway receded from the attainment of this high ideal.

At each station, crowds were gathered to watch and as the train steamed into Kinnerley, the 82-year old Mr. Reeves was waiting on the platform in welcome, wearing the uniform with which he had brought in the last "Potteries" train more than thirty years before; he joined the party for the trip to Llanymynech, which was punctuated only by a stop at Maesbrook. At the terminus a half hour stop was made; the Chairman of the Llanymynech Parish Council, Mr. Kemble, read from the platform a short address of greeting, in which Mr. Stephens was thanked for employing local labour whenever possible. The party then refreshed themselves upon sandwiches, the Loving Cup went round, and to the accompaniment of fog signals and cheers the return journey was commenced; at Kinnerley, the party detrained once more to inspect the locomotive shed and shops and the train finally reached Abbey Foregate at about 3.30 p.m.; the weather had been ideal and this most auspicious start augured well for the line.

Any complacency was soon to be rudely shattered; on the following day (Good Friday) the line was first opened to the public and many hundreds of country folk took advantage of the opportunity of spending a day or a weekend in Shrewsbury. They were more (or less, perhaps, in some cases) fortunate than those who turned up to catch the early morning train to Shrewsbury on the Saturday, for the locomotive *Hesperus* and three coaches were derailed when negotiating the sharp curve by Red Hill; fortunately no one was hurt, though the accident occurred on an embankment and the track was damaged for some distance, for there was no overturning. The tender, oddly enough, never left the rails. The L.N.W.R. breakdown unit was summoned from Shrewsbury and the gang worked all night, the line finally being cleared by Sunday afternoon; many visitors to Shrewsbury who had hoped to return home on Saturday were stranded and had to get back as best they could by motor or brake.

The Company, however, did not allow themselves to be deterred by this unfortunate occurrence(which indeed might have ended much more tragically) and the S. & M. settled down to the humdrum routine of everyday operation. From the passenger angle, there were three trains (timed to meet Cambrian trains at Llanymynech) each way on weekdays, with two extras on Thursdays and Saturdays, and two each way on Sundays. The journey occupied one hour, or a few minutes more; which schedule, it will be observed was slower than that obtaining in 1866, due to the track's being single, the 25 m.p.h. speed restriction and last—but, alas, certainly not least—to the

---

* *The Shrewsbury Chronicle*, April 21st, 1911.

47

Kinnerley Station looking east, April 23rd 1939. The origins of the coaching stock at the bay platform are respectively (l to r): Midland, Midland, N.S.R., N.S.R., L.&S.W.R., L.&S.W.R. The four wheel passenger brake van on the main line is also Midland Railway; the end windows are not original. (L. W. Perkins)

The road bridge at Melverley, looking south. The station building and wagons in the siding can be seen to the right and left beyond the bridge. April 23rd 1939. (E. S. Tonks)

fact that all trains were "mixed", with all the delays and discomforts of intermediate shunting operations. Cheap half-day tickets were issued on three days a week to many stations, and by arrangement with the Cambrian Railways, excursion tickets from Oswestry to various points on the S. & M.R.; market tickets to Oswestry and Shrewsbury were also introduced.

The Criggion branch was still closed to traffic pending the completion of the new Melverley viaduct*; this was tackled energetically, however, and on Wednesday, February 21st, 1912 the branch was opened for parcels, goods and mineral traffic, whilst the promise of catering later for passengers was fulfilled in August of the same year. The Pyx Granite Company, which expected to benefit considerably from the railway (their output having hitherto to be transported expensively by road), had erected a crushing plant at the quarry; hopes were also entertained of opening quarries at Belan Rock, and a syndicate had been formed to exploit coal in the Crew Green district.

The Superintendent of passenger and goods traffic was Mr. J. L. White, for many years an Inspector on the L. & N.W.R. at Shrewsbury; the chief of the Rates and Fares Department was Mr. John Stockdill, who came from the L. & S.W.R.; and Mr. A. J. Matthew, late of the S.E. & C.R. accounting department, became Audit Accountant. The beneficent rule of these gentlemen under the guidance of Colonel Stephens resulted, in the first six months of operation, in a profit of £404-4-6; after payment of debenture and share holders, the reserve account came in for the princely sum of 13/3d. 42,000 passengers and 17,500 tons of goods had been moved by the mixed trains (there were no purely passenger or goods trains), a substantial increase on the corresponding "Potteries" figures. The same report revealed that only 100 shares had so far been issued (the Act, incidentally, provided that no more than 100 could be issued without Shropshire Railways' consent).

For 1912, in which year a net profit of £463 was made, the timetable was somewhat modified, having three trains on the main line from Shrewsbury to Llanymynech and four in the reverse direction, with an additional train from Kinnerley to Shrewsbury and back on Thursdays and Saturdays; there were three branch trains with another early morning train Wednesdays and Saturdays only, whilst the Sunday service consisted of two main line trains and one branch train each way.

Except for alterations in the departure times and an average increase of 15 minutes on the overall schedule, there were no substantial changes in the main line service during the succeeding twenty years. Unhappily, the same could not be said of the passenger traffic receipts, which fell off steadily;

---

* This is believed to have been constructed of girders from the disused side of Shrawardine viaduct.

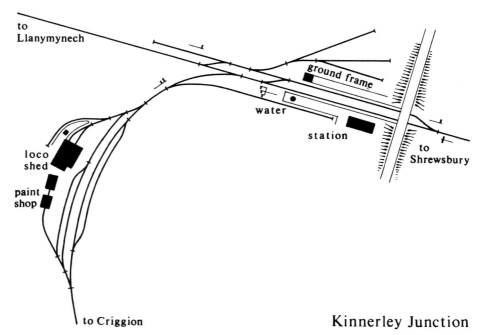

to Llanymynech

ground frame

water

station

to Shrewsbury

loco shed

paint shop

to Criggion

Kinnerley Junction

the S. & M. were in a less fortunate position than the "Potteries" in this respect since, in addition to the initial sparseness of the population in the country served, there was the ever-growing menace of competing road transport. The S. & M. had more stations, true, but a glance at the map will show that the tiny hamlets they represented could hardly be expected to provide enough revenue even to pay for the platforms. Moreover, in too many cases the stations were merely at road crossings some distance from the actual village, themselves directly served by the country bus. Again, though certain purely mineral trains were soon introduced, all the passenger working were still mixed and the seemingly interminable shunting operations—amusing enough to the railway enthusiast—were tedious to the would-be regular passenger.

In spite of these handicaps, the S. & M. struggled valiantly to retain its passenger traffic: the country lanes were plentifully signposted with directions to the nearest station; some of the rolling stock appeared in a livery of light blue with red and yellow figures; at Crew Green for Coedway and Alberbury (to use the full title) camping huts and boats were to be hired; whilst all the timetables carried the exhortation "support the local line", often coupled with invidious references to the use of imported fuel.

In an effort to cater more economically for the reduced number of

50

passengers and to attract more by segregating passenger and freight workings, Colonel Stephens introduced in the early twenties two of his celebrated
back-to-back railcar sets, a Ford pair and a Wolseley-Siddeley pair. These
were ingenious and indeed cheap to run, but beyond that, virtues had they
none. The writer was not able to have such a trip on this line but did have
the pleasure (an emotion, be it emphasised, derived entirely from the novelty
of the experience) of riding the similar Ford and Shefflex contrivances
of the K. & E.S.R.; the seats were hard, the body incredibly vibratory and
noisy, and progress slow, and as an alternative to the steam trains even of
S. & M. standard the railcars were a hopeless failure from the start, whatever economies in operating expenditure they may have effected. They
looked ridiculous, too, and made the staff apologetic and self-conscious in
face of the often unprintable comments of passengers; as one driver succ-
inctly put it: "Why, it ain't a bloody railway any more; just a broken-down
bus on rails." They were short-lived on the S. & M. but not before they
had added their quota of damage to the passenger receipts; the Criggion
service (already reduced to two trains each way) was the first to suffer the
axe, and from September 1928 branch trains ran only on Saturdays; and as
from October 1932 trains terminated at Melverley, the viaduct beyond being
considered too frail for intensive passenger working. By this time, steam
working was again the order of the day and the Melverley-Kinnerley train
consisted of one coach propelled in front of the engine, there being no loop
at Melverley.

Ford Railcar set at Llanymynech.   (*L. & G.R.P. 4384*)

51

Kinnerley loco shed in the rain, with the relics lined up in the siding: GAZELLE, *tramcar, Ford rail-lorry and railcar, tender off "Ilfracombe goods". Cab by water tower, boiler onleft. "Collier" in shed.* (L. & G.R.P. 6763)

Kinnerley station, looking west with one of the "Collier" locomotives shunting.
(Tennent, Halesowen)

The last advertised timetable, operating from October 1932, consisted of a morning trip Kinnerley-Llanymynech-Shrewsbury-Kinnerley, and a similar trip in the afternoon, the latter running at a later time on Mondays and Wednesdays; on Saturdays there was also a midday train from Kinnerley to Llanymynech and back, and the two branch trains. Sunday services had, of course, ceased long ago. It was soon clear that the downward trend of the passenger traffic was not to be balanced by economies and on November 6th, 1933, almost without notice, passenger services on the S. & M. were withdrawn for good.* For a few more years, however, excursion trains were still run at Bank Holiday periods and considerable numbers of people travelled by train to the various hilly beauty spots within convenient reach of the line—the Breiddens, Nesscliff Hill and Llanymynech Hill. An added and well-patronized attraction on these trips was the presence of the ex-L.S.W.R. Royal Saloon, for travelling in which a supplementary fare was charged. Finally, even this meagre survival ceased and the only way of travelling on the S. & M. from 1937 onwards was by special hire of *Gazelle* and its attendant coach; a few local pleasure parties availed themselves of this opportunity and at least one band of light railway enthusiasts—the Birmingham Locomotive Club—ran a trip (and a more enjoyable day's outing there never was) on Sunday, April 23rd, 1939, and another on the following Sunday to accommodate the overflow. What memories—nostalgic yet happy—their recollection calls up! The grass-grown track: the human chain of buckets from a pond at Criggion to help Driver Owen to fill the tiny water tank through an inverted taper top can with the bottom sawn off: the two cyclists who nearly fell off their machines at Llandrinio Road at the incredible sight of the "wooden engine" and coach and its complement of laughing fellows: the even greater consternation of the fisherman who was all but decapitated when he stuck his head through Melverley viaduct on hearing the unprecedented Sunday train; the fun of chasing sheep off the line; and the pathetic notice at Llanymynech announcing " Frequent trains and cheap fares to Shrewsbury." An unforgettable experience; nor can its like ever be again.

The encroachment of road transport was by no means confined to the passenger side of operation and throughout the foregoing period many small sources of freight revenue had been lost by the railway; by the middle thirties nothing was left save a little parcels and agricultural traffic and the granite from the Breidden Hills quarry of the British Quarrying Co., which had taken over the Pyx Granite Co's works. The geographically remote terminus at Criggion thus became the most important point of the system, and there is not the slightest doubt that the S. & M. would have closed long before but for the stone traffic; even so, this was small enough (about 250 tons daily) calling for one trip each weekday, leaving Kinnerley at

---

* The L.M.S. Timetables showed a daily train after this date, but this was actually the goods working.

GAZELLE *and Inspection Car at Criggion. April 30th 1939.* (*W. A. Camwell*)

6 a.m. for Criggion, thence to Shrewsbury and back to Kinnerley, or Llany-mynech if required, finishing about midday; in winter, the Shrewsbury trip was performed first. By this slender thread hung the life of the Shrop-shire & Montgomeryshire.

The declining fortunes of the railway were well mirrored in the condition of the equipment; except in the vicinity of important stations, the permanent way became carpeted with grass and weeds, the stations (from all of which except Shrewsbury, Kinnerley, Meole Brace, and Ford staff had been with-drawn many years before) fell into disrepair and some were boarded up; whilst Kinnerley assumed the appearance of a neglected museum. It is doubtful if the S. & M. ever required more than two engines in steam at any one time; in the twenties, one steam loco and a railcar sufficed, and in later years the normal traffic called for the labours of but one engine for half a day. In Colonel Stephens' time, no engines were ever scrapped but (to use an exceedingly apposit cliche) merely faded away; it was also the Colonel's custom to provide his railways with a locomotive stock considerably in excess of everyday needs and whether his motive was caution or optimism, the result was usually the same—the cannibalization of some engines to keep the others moving. The result was to give to Kinnerley and its counter-parts at Rolvenden, Selsey etc., an air of decaying munificence which most light railway enthusiasts found more attractive than streamlined efficiency.

HESPERUS *on the daily goods at Kinnerley, April 11th 1938.* (*R. K. Cope*)

Colonel Stephens died in 1931 and his offices at Salford Terrace, Tonbridge, were taken over by the newly-appointed Manager and Receiver, Mr. W. H. Austen. Mr. Austin's policy differed from the late Colonel's; instead of a large loco stock, of which half were just kept going by depredations on the remainder, he instituted a policy of restoring two or three engines to good working order, including the application of a long overdue coat of paint, and scrapping the rest. A similar principle was applied to the rolling stock, with the net result of a railway more efficiently equipped but offering less to the railway fan poking about among the thistles in Kinnerley yard. These changes took time, however, and throughout its independent existence the S. & M. retained in good measure that ingenuous charm characteristic of hard-up light railways. A visit to Kinnerley was a joy; standing on the yellow-painted bridge over the station in the drowsy stillness of a summer day, there was nothing to mar the scene; from the direction of the engine shed an occasional clink, and a thin column of steam rising from the workshop; the many-vaned windpump; and the old rolling stock, discarded and forlorn, parts half-hidden in a green cloak of brambles and providing shelter for nesting Thrushes and Blackbirds; even the gentle ministrations of the one ancient engine in steam were as much in keeping with the rural surroundings as a haycart. It was only too obvious, however, that the railway could not survive indefinitely in this state of affairs and many enthusiasts cast anxious eyes on the S. & M., expecting its closure; there

"Collier" 8236 at Llanymynech with the ex L.&S.W.R. Royal Saloon and two ex-Midland vehicles.   *(Collection M. Rhodes)*

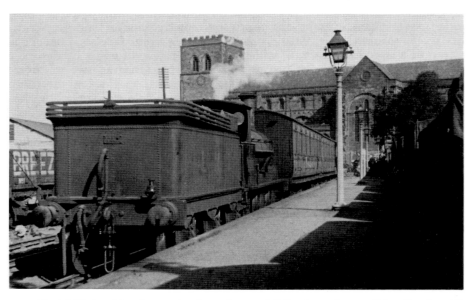

Ex-LMS 8236 heads a train of two former Midland coaches at Abbey Station. The storage vans referred to in the text appear on the right.   *(Collection M. Rhodes)*

56

were grounds for their fears, too, in the diminishing stone traffic, but somehow the line was kept open.

At the outbreak of war, the S. & M. was one of the few smaller systems to come under Government control and in consequence received a guaranteed annual profit of £1. But to outward appearances the railway was unaltered; and so it continued—Locomotive No. 2 handling nearly all the traffic—until 1941, when the War Department, which in their search for suitable sites for depots and defence works seem to have had a predilection for little-used or abandoned railways, requisitioned the whole of the main line as the trunk of a vast network of stores; the story of the railway then took a very different turn, as we shall see in a later section.

Locomotives and rolling stock

The Shropshire and Montgomeryshire Railway possessed at one period or another fourteen steam locomotives, excluding various engines loaned on occasion; of these, six were tender locos, an unusually high proportion for a light railway, but the S. & M. has always been essentially a tender engine line. At the time of opening, the locomotive stock consisted of a very interesting, if motley, collection of six engines, all bearing names and numbered 1 to 6 in alphabetical order; whether by design or chance is not known.

No. 1, *Gazelle*, was the most remarkable of these and can well claim to be the most famous standard gauge light railway locomotive in the British Isles, having risen to almost national prominence. This unique machine was built in 1893 by the engineering firm of Alfred Dodman & Co. at their Highgate Works, Kings Lynn, to the special order of Mr. William Burkitt, who combined the qualifications of an influential business man (he was twice Mayor of Lynn) with a rare love of locomotives; being on good terms with the G.E.R. and the M. & G.N., he had his own loco built and used it on business trips from time to time, on one occasion going so far as Chesterfield (Market Place). At other times it was stabled at Lynn.

It was one of the smallest standard gauge locomotives in the world, and was built as a 2-2-2WT, with seats for four people at the rear of the cab. Leading dimensions were: driving wheels — 3' 9"; other wheels — 2' 3"; wheelbase — 10' 6"; cylinders — 4" x 9"; centre line of boiler — 3' 11" above rail level; height to top of chimney — 7' 9"; length over buffers — 17' 2"; weight — 5 tons 6 cwt. To reduce noise, the loco had Mansell wheels with polished teak segments. She was offered for sale by Messrs. T. W. Ward in 1910 and was purchased by Colonel Stephens in February 1911, to serve as an Engineer's Inspection Unit. He sent her to W. G. Bagnall Ltd. of Stafford for rebuilding as a 0-4-2WT by providing a centre pair of wheels of design similar to the leading pair. *Gazelle* returned to Kinnerley in July 1911 but went back to Bagnalls for the cab and rear seating accommodation to be covered in, and a step added. Thus altered, the engine ran for many years

57

GAZELLE *as received by the S. & M. in 1911, before rebuilding as a 0-4-2WT.*
(Authors Collection)

GAZELLE *nameplate and S. & M. plate. Photographed when derelict at Kinnerley.*
July 11th 1933. *(P. W. Robinson)*

GAZELLE *as restored in 1937, with cab and rear compartment. April 30th 1939.*
(P. W. Robinson)

GAZELLE *at Kinnerley. Note the rails around the rear roof which were removed during the 1937 rebuild.* *(Real Photographs Co. Ltd.)*

the bulk of the branch traffic, in company with an ex-L.C.C. horse tramcar (numbered 16 in the coach stock).*

The reduction of the Criggion branch passenger service from independance to a mere supernumerary trip to the main line run rendered the loco redundant and in the early thirties it was discarded along with the tramcar and lay derelict behind the water tower at Kinnerley until 1936; by which time much of the boiler casing and mountings had disappeared and the whole engine in a deplorable state. In this year, however, Mr. W. H. Austen (who had succeeded the late Colonel Stephens as Engineer and General Manager) decided to restore *Gazelle* to its original condition and status, i.e., the operation of inspection trains, and so it was taken into the repair shop, the missing parts ferreted out from amongst the nettles and in June 1937 emerged resplendant in an olive green livery with black and white edging, red buffer beams and connecting rods, brown frame, green wheels, grey buffers and couplings and polished safety valve, dome and nameplates; on the cab sheets was painted " S. & M.R. 1."

A body from the Wolesley-Siddeley railcar set,† mounted on the underframe from the erstwhile tramcar, completed the train and this combination operated for inspection purposes and private parties until the Military took over the line. Even then, however, *Gazelle* had not outlived its usefulness; the opening up of the various yards called for considerable track reorganization and it was *Gazelle's* duty to run over the main line before the first train left Kinnerley, to ensure that all the points were correctly set and (it is said) to counteract the activities of any saboteurs. Thus employed, *Gazelle* (repainted light green with black framing and red motion) outlived all the other S. & M. stock, but eventually its job was allotted to a Wickham petrol car and once again it was taken out of service (its last public appearance being in 1945 in connection with a National Savings Campaign) and stored at the end of the siding by the pond. While here, a "Dean" was allowed to run into it and damage the cab; the funnel, too, had lost a portion of the top on a trip to Criggion and the stump had been sawn off level and a crude lip riveted on, but apart from these the engine was complete and Mr. Calder resolved that she be preserved by the W.D. In May 1950 she became nominally BR(WR) property but the following month was (again on paper) transferred to the W.D. on permanent loan. She was sent to Longmoor for preservation in June 1950; repainted and repaired she was shown on "Open Days" until the closure of the Depot in 1970. At the time of writing she is still at Longmoor but she may be moved to Aldershot.

* References: *Railway Engineer*, August 1893, p. 257; *The Locomotive*, Vol. 6, 1901, p.80; *Steam engine builders of Norfolk*, Ronald H. Clark, Augustine Steward Press, 1948; *Industrial Railway Record*, Industrial Railway Society, No. 24, April 1969, p.84; No. 26, August 1969, p.126; No. 28, December 1969, p.178.

† It has, however, been reported that this body came from one of the Selsey cars.

No. 2, *Hecate*, was a more orthodox type of locomotive (frankly *Gazelle* always looked as if it belonged to a fairground) but from the historical point of view even more interesting. It was purchased in September 1911 from the Griff Colliery Company, sent to Messrs. Bagnall's for reconditioning—including the fitment of a vacuum brake—and put to work on the S. & M. Its early history is shrouded in mystery, but examination of the bar frames for the coupled wheels and of the valve gear, together with information supplied by its former owners, suggests that it was built by Bury Curtis & Kennedy of Liverpool in the 1840's, for the Shrewsbury & Hereford Railway as a 0-4-0 tender engine, subsequently passing into the hands of the L. & N.W.R. and sold by them in 1871 to the Griff Colliery Co. When received by the S. & M., it had been converted to a 0-4-2ST with raised firebox and a dome on the centre of the barrel, but retained its bar frames and round section connecting rods, working beneath the front coupled axle to the downward sloping cylinders. At Griff it had borne the name *Crewe*, but at Kinnerley, *Hecate* was painted on the tank sides inside the oval carrying the Company's title; in 1916 the name was changed to *Severn*, brass plates being cast for this purpose; at a later date the squat funnel was replaced by a taller tapering one. At first it worked mixed trains on the Criggion branch, but in the early twenties was applied to stone traffic solely and worked thus until about 1929. Officially withdrawn in 1931, it lay dere-lict in Kinnerley yard — gradually disintegrating as parts were used to repair the other relics — until April 1937, when the remains were broken up on the spot by a local scrap merchant and all that survived was one buffer beam.

No. 3, *Hesperus*, was a 0-6-0 tender engine purchased in January 1911 from the London & South Western Railway, on which line it was one of a series of eight such engines supplied by Messrs. Beyer Peacock to Mr. W. G. Beattie's design, for use on the steeply-graded but somewhat lightly laid Ilfracombe line; the class naturally became known as the "Ilfracombe goods" and their characteristics made them ideal for light railway operation, as Colonel Stephens was quick to appreciate, for he purchased no less than six of them, three of which came to the S. & M. *Hesperus* was built by Beyer Peacock & Co., Manchester, in April 1875 (Maker's No. 1517), and rebuilt at Nine Elms in June 1888 with an Adams boiler; originally L.S.W. 324, it was placed on the duplicate list in November 1900 as No. 0324, and withdrawn in August 1910.*

On the S. & M. it was fitted with vacuum brake and steam heating for passenger work and was painted olive green with light green lining, with the name painted on the centre splasher, the number in black on the front buffer beam and in gold on the back and the letters S. & M.Rly. on the tender; later, cast nameplates were provided. A query may appositely be

---

* "Clausentum", the *Railway Observer*, June 1943.

SEVERN *formerly* HECATE. *A much rebuilt Bury locomotive.* (*F. Moore. 5229*)

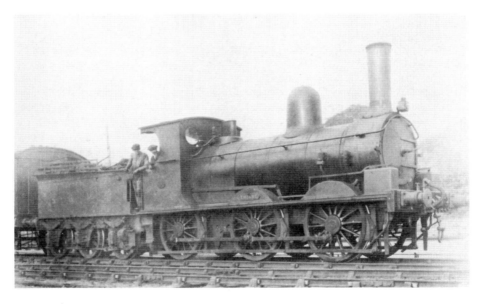

One of the S. & M.R.'s "Ilfracombe goods"—THISBE. (*Authors Collection*)

put here; why was Colonel Stephens so fond of this name, which was perpetuated on no less than four of his standard gauge lines? As already recorded, to *Hesperus* went the distinction of working the train for the ceremonial opening and it handled much of the passenger services in the earlier days of the Company's existence; it was given a new lease of life after a repair at Crewe in 1919, but with the arrival of additional locomotive power in the twenties it was relegated to the position of spare and was indeed "stored" (a process involving the placing of a tile over the funnel) at Kinnerley for some years. During Mr. Austen's regime, however (about 1934), it was repaired and put back into service, thus outlasting the other engines of the class on this line; in the late thirties, the "Colliers" handled most of the traffic, *Hesperus* usually working on Mondays when the stone traffic was light. As late as June 1938 it was overhauled and the smokebox and funnel painted black (the original livery was by now unidentifiable under a multiplicity of patches) and though far beyond its prime, was still capable of a day's work; but it did not long survive W.D. control and in September 1941 *Hesperus*, the last surviving "Ilfracombe goods", was ignominiously towed to Abbey Foregate yard and in November broken up on the spot, a forgotten engine.

No. 4, *Morous*, was a six-coupled saddle tank built by Manning Wardle & Co., Leeds in 1866 (Works' No. 178) and was one of a pair supplied to

MOROUS *outside Kinnerley shed.*   *(Real Photographs Co. Ltd.)*

Messrs. T. B. Crampton for the construction of the East & West Junction Railway; this particular engine was sent to Fenny Compton (the sister engine M.W.177 going to Stratford) and became E. & W.J.R. No. 1, was rebuilt by this Company in 1896, becoming S. & M.J.R. No. 1 in 1909; Colonel Stephens purchased it in 1910. *Morous* had a red livery, lined in yellow and black, with the name painted in yellow inside an oval bearing the words "Shropshire and Montgomeryshire Railway"; later, brass nameplates were fitted a few inches above the original name, which still appeared. The writer has been unable to trace the origin or meaning of this odd name: but up to the outbreak of war there were preserved at Kinnerley a pair of plates *Morus*, presumably cast in error arising from the local pronunciation.

This engine does not appear to have been very much used (small tank engines never found much favour on the S. & M.) and in November 1924 it was transferred to the West Sussex Railway, another of Colonel Stephens' lines and one which was practically monopolized by contractor's type locomotives. On this system, *Morous* was well-liked and much used on mixed trains, being vacuum fitted, and along with the incapacitated *Selsey* remained in Selsey shed after the closure of the line in January 1935 : it became "Lot 323" at the auction of the West Sussex Railway equipment in June 1936 and rumour states that it was put into steam for handling the demolition train; whether the latter is true or merely wishful thinking,

MOROUS *in the Selsey shed after the auction of the West Sussex Railway equipment in June 1936. Still in S. & M. livery.* (R. W. Kidner)

support is lent by the fact that the track removal did start from the Chichester end. This work over, *Morous* was cut up in the autumn of 1936, retaining its S. & M. livery to the end.

No. 5, *Pyramus* and No. 6, *Thisbe* shared the distinction of being the only new engines the railway ever possessed; they were 0-6-2 side tanks of handsome appearance, built by Messrs. Hawthorn, Leslie & Co. in 1911 (Maker's Nos. 2878/9) to Colonel Stephens' drawings. Like *Morous*, the names were first painted inside the S. & M. "totem" on the tank sides, but later, cast plates were substituted. They were fitted with Stephenson's valve gear, combination steam and vacuum brake, stovepipe funnels and had rapid acceleration; unfortunately, they proved somewhat too heavy for the track and were both sold at the end of 1914 to the Woolmer Instructional Military Railway, becoming W.D. Nos. 85 and 84, respectively.

Later both went on to Kinmel Park Camp, near Rhyl, and appear to have been disposed of after the war. One was sold to James Clements of Cardiff, resold by him to the Mersey Docks & Harbour Board, where it became No. 34. As their Nos. 33 and 35 were acquired in 1922, it may be assumed that No. 34 came in that year also. She was sold by 1927 and was acquired about 1930 by Nunnery Colliery Co. Ltd. near Sheffield; she remained there under N.C.B. ownership and was scrapped by C. F. Booth Ltd. on site in April 1962. This locomotive is generally believed to have been *Thisbe*, but this identification has been queried, and an element of doubt remains. Quite possibly some exchange of parts with the sister engine led to this confusion. Another version quotes *Pyramus* and *Thisbe* as becoming 84 and 85 respectively, i.e. the "correct" way round, and that *Thisbe* as 85 was scrapped in 1926. This would make the surviving Nunnery Colliery locomotive as the former *Pyramus*.

The 0-6-2T's were replaced by two more "Ilfracombe Goods", which took the same numbers and names. No. 5 was built in June 1874 (B.P's No. 1428) as L.S.W. 300, rebuilt December 1890 with Adams boiler, renumbered 0300 in December 1900 and withdrawn from service in January 1914, to be purchased by the S. & M. in November 1914. No. 6 was even earlier, B.P's No. 1209 of February 1873, L.S.W.283; rebuilt June 1888, renumbered 0283 October 1899 and withdrawn January 1914; purchased by the S. & M. in May 1916. Both engines were given a smart blue livery with red lining and had brass nameplates (these plates had curved ends, while those of *Hesperus* were rectangular) on the splashers, the number in black on the buffer beam and letters S. & M.R. on the tender; like the other engines, Nos. 1-4, they also carried on the cabside the Company's oval plate with the wording "Shropshire and Montgomeryshire Railway Company". They were fitted with vacuum brake, but on the arrival of the railcars spent most of their time on freight workings. *Pyramus* was the less successful of the

PYRAMUS. *The makers photograph, with original painted name.*
(*F. Moore. 4076*)

THISBE *at Llanymynech with nameplate as fitted later. The coaching stock is of*
L. & S.W.R. origin. (*P. W. Robinson*)

two and was taken out of service in the late twenties and allowed to rust away for some years, finally being cup up in 1932. Its boiler, however, was in fair condition and was given to *Thisbe*, which ran until 1935, handling passenger trains on occasion; *Thisbe* then lay derelict at Kinnerley for two years and was cut up on May 18th, 1937 by a local scrap dealer.

The next arrival, *No. 7 Hecate* was one of the ex-L.B.S.C. A.1. "Terrier" 0-6-0 Tanks, another class favoured by Colonel Stephens; this was built in 1880 as No. 81, *Beulah*, later becoming 681, and was purchased from H.M. Docks at Invergordon (where it had worked during the war) in August 1921. Two more of the class were purchased from the War Stores Disposals Board at Dalmuir in November 1923; these were 638 (built as 38, *Millwall*, in 1878) and 683 (built as 83, *Earlswood* in 1880) and became No. 8, *Dido* and No. 9, *Daphne*, respectively. Numbers were painted in black on the buffer beams. Strangely, the "Terriers" were not very successful on the S. & M. and the first two were taken out of service as early as 1930; they were subsequently (circa 1932) dismantled and the boilers mounted on piles by the Criggion branch near the shed, where they stood until about 1934. Parts were sent to Rolvenden for the rebuilding of K. & E.S.R. No. 3, *Bodiam*, which almost certainly acquired the tanks from *Hecate*. Bit by bit the other remains vanished (the frames had gone by 1937), the last trace being one of *Dido's* tanks,

The end of a veteran. The later THISBE *being cut up in Kinnerley yard, May 18th 1937.* (*E. S. Tonks*)

67

A Brighton "Terrier" in Shropshire — **HECATE** *at Kinnerley.*
(Real Photographs Co. Ltd.)

DAPHNE *in the round-topped shed at Kinnerley, September 1st 1937.*
(R. K. Cope)

68

complete nameplate, which lingered until 1939. *Daphne* was more fortunate and, though withdrawn from service only a short while after her sister engines, was stored in the corrugated iron erection behind the running shed and there remained until December 1938, when she was purchased by the Southern Railway as a source of spares for their "Terriers" still in service. The Southern did not make much use of their bargain, however, and *Daphne* is still to be seen at Eastleigh in the company of a few other delightful relics, weatherbeaten and stripped of plates but retaining a copper-capped funnel, until cut up March 1949.

No more steam locos were purchased during Colonel Stephens' time but, shortly after Mr. Austen took office, three more 0-6-0's — all ex-L.N.W.R. "Colliers"—were obtained to replace the three "Ilfracombe Goods", all of which were by then a bit tottery; the historical background of these engines is given in the following table:—

Built	Crewe Wks. No.	Orig. L.N.W. No.	Later No. and date	L.M.S. No.	Date of purchase
Dec. 1874	1869	2167	3563 (Jan. 1910)	8108	Mar. 1930
Dec. 1879	2333	155	3045 (Dec. 1916)	8182	June 1931
June 1881	2459	2422	3575 (Feb. 1910)	8236	Aug. 1932

8236. The appearance of this engine altered little on the S. & M.R. and the photograph could possibly have been taken at many places on the L.M.S. in the previous decade. It was, in fact, taken at Kinnerley on April 23rd 1939. (*L. W. Perkins*)

No. 2 (ex 8108) in green livery, shunting at Kinnerley, August 24th 1939.

(R. K. Cope)

These engines retained their former L.M.S. livery, i.e. black with yellow numbers on the cabsides and letters on the tender (and, in the case of 8236, a cast numberplate on the smokebox door) and between them handled practically all the shrinking traffic over the next ten years. The oldest, 8108, was the first to fall due for a heavy repair, a task carried out in the running shed over a period of 18 months in the leisurely fashion imposed by the financial limitations of the Company.* The boiler was retubed, a new smokebox obtained from Crewe Works, where the wheels were sent for re-turning, and finally in May 1939 there appeared a "Collier" replendent as no other of its class had been for perhaps half a century, in the original "Southern" green lined with black edged with white, with a large figure 2 in yellow on the cab panels and the letters S. & M. on the tender, also in yellow. It is believed that plans were in mind to recondition the other two "Colliers" as S. & M. Nos. 4 and 5, but the outbreak of war prevented their fruition; instead, the W.D. painted all three in "camouflage" green, later embellished with red lining, with the numbers in white on the cabsides, No. 2 reverting to 8108. These three worked under W.D. control until the traffic fell off, when they were withdrawn en bloc (in 1946) and taken to the new S. & M. graveyard at Hookagate; they became B.R.(W.R.) property

---

* The gentleman in charge of the loco. shed was — believe it or not — a Mr Funnell.

in May 1950 and were sent to Swindon for scrap last being observed on the "dump" in September 1950.

Other engines have worked on the S. & M. from time to time; including *Walton Park*, a 0-6-0ST, Hudswell Clarke's 823 of 1908, which had been No. 4 on the Weston, Clevedon & Portishead Railway; this came to Kinnerley in 1913 and in 1916 was transferred to the East Kent Railway, where it became their No. 2. Again sold in 1943, it became after a number of vicissitudes *Churchill* of the Purfleet Deep Water Wharf & Storage Co. In 1927 the "Sentinel" locomotive building firm of Shrewsbury conducted trials on the S. & M. with the use of their Super-Sentinel vertical boilered locomotives (working passenger turns) and steam railcars. Occasionally a "DX" 0-6-0 was borrowed from Shrewsbury L.N.W. shed, e.g. 3128 in 1921.

In addition to the steam stock, the S. & M. possessed at one time two back-to-back railcar sets of the type introduced by Colonel Stephens on other of his light railways, the K. & E.S.R. and the West Sussex Railway: in these sets, both cars had engines, but that of the rear car was disengaged, according to the direction of travel. The first (in fact the pioneer) and more successful of these was a Ford three-car set—the middle car being simply a trailer—which was built by Messrs' Edmunds of Thetford in 1923, and seated 19 persons per car, with 24 in the trailer; each car was 17′ 4″ long, and had 2′ 6″ wheels. The second was a Wolseley-Siddeley pair purchased later in the same year, and operated with a light open truck between. These railcars are so well known that there is little need to dwell on their characteristics (not to say eccentricities); they were cheap to run and required no turntables, but were not calculated to inspire confidence in the travelling public. They were deficient in adhesion and in wet weather frequently stalled on the bank out of Abbey. The Wolseley-Siddeley car was taken out of service about 1929 and scrapped about 1935 (some of the seats surviving on station platforms) but the Ford pair (painted blue) lasted longer and as a derelict survived until 1941. There was in addition a Ford-engined single unit goods vehicle that operated with half of the other Ford set, i.e. forming a "mixed" railcar. The history of this machine is unknown but Mr. Richard H. Stuart, who provided a photograph of it taken in June 1928, and who is a "Model T" enthusiast, states that it is of the type built 1919-25, and the cab suggests the early part of the period; like its antecedents, its fate is unknown.

The rolling stock of the S. & M. was all purchased second-hand and at first consisted of four bogie coaches (two 7-compartment thirds and two with 2 firsts, 3 thirds and guards compartment) and two four-wheeled brakevans, all from the Midland Railway: these were in full Midland livery of crimson with gold lettering and looked very smart. Later in 1911, six four-wheeled coaches were obtained from the Plymouth, Devonport & South-Western Junction Rly; these were all ex-L.S.W.R. stock and com-

Ford combined railcar and goods vehicle set at Llanymynech. About 1926.
(J. A. Hignett, Oswestry)

prised three 5-comps. thirds, one first and third composite, and two with 3 thirds and guard's compt. These, together with three ex-N.S.R. four-wheeled coaches obtained a few years later, became Nos. 1-15, while the L.C.C. tramcar purchased for use in combination with *Gazelle* became No. 16. About the same time also was purchased the showpiece of the rolling stock, a four-wheeled vehicle from the Victorian Royal Train built by the L.S.W.R. in 1848; three coaches of this train were purchased by the P.D. & S.W.J.R. for use as a set and two later came into Colonel Stephen's hands for use on the K. & E.S.R. and the S. & M. The last was numbered 1A and painted blue; it was well sprung and comfortably furnished and even the W.D. did not disdain to use it. In January 1953 W. G. Cross reported it as "recently repainted blue and now part of the breakdown train". She was sent to Longmoor in December 1953 and scrapped there. The other coaches were painted ultramarine with vermilion ends (*The Locomotive*, 1923, p.128) but later on a buff livery was adopted; these coaches and one late arrival of G.E.R. origin transferred from the K. & E.S.R. were all laid aside after the cessation of public services, an ex-Midland vehicle being the last in use; Nos. 1, 2 (both brake vans), 3, 6, 8-10, 12-14 were kept in the bay platform and yard at Kinnerley, and two others in the bay at Abbey, the rest having been scrapped already. One of the ex-N.S.R. coaches was repaired in 1940 but was scrapped along with the rest of the passenger stock at Kinnerley by the W.D. Van 2 is stated by the W.D. to have been at Nesscliff and was converted to a platelayers' cabin. The two at Abbey were 7 (coach) and 17 (brake third) according to the W.D., and there was also a passenger van, 18. 7 was converted to a flat wagon in 1941 and scrapped in 1943 or 1945 (W.D. records quote both dates); 17 and 18 were scrapped in 1952.

The goods rolling stock was all secondhand, too, even including one ex-P.S. & N.W.R. open wagon which survived as a wreck until 1936; beyond this and a travelling crane there was nothing remarkable; the crane was a 4-wheeled 5-ton hand operated one with wooden jib; she was reported by the W.D. to be at Shrawardine, withdrawn October 1952 and sent to Swindon July 1953. In 1912 the total number of vehicles was returned as 32, but this number was augmented later to nearer the 50 mark; few of this number were in use after 1930 and a year or so later Kinnerley possessed but a collection of rusty underframes and rotting woodwork. The stone traffic was conveyed in the B.Q. Co's private wagons. The goods stock was painted grey with white lettering (sometimes the Company's name being in full) and each wagon boasted a plate with the words "Shropshire & Montgomery-shire Railway Company" but rectangular instead of oval as used on the

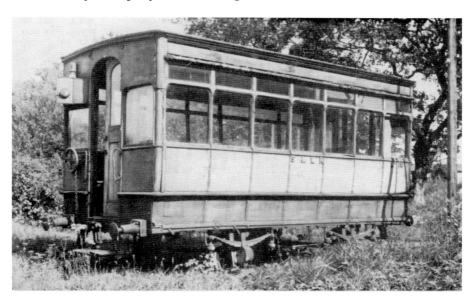

The ex-London County Council horse tramcar at Kinnerley on September 11th 1933.
(P. W. Robinson)

locos. It may be mentioned here that Guards Vans for goods trains were not used, a disc bearing "L.V." being hung on the coupling hook of the last vehicle.

W.D. records give the following distribution of goods vehicles, at a date unknown but probably when they assumed control; all were out of use. Llanymynech — Timber: 1, 24-6. High-sided: 44, 49. Low-sided: 3. Criggion — single bolster: 2, 3. Shrawardine — Runners: 9, 10. Horse box: 15. All these are recorded as scrapped 1948. Fifteen vehicles were used at Abbey station for storing grain and basic slag, for which purpose they were covered with tarpaulins. W.D. quote these as Box Wagons 15-18, 20-24 and Cattle Wagons 30-3, 42, 43. The box wagons were converted to platelayers, cabins in 1952 and in 1959 six were still in use at Edgebold station, Horton Lane, Shoot Hill, Kinnerley Triangle and Kinnerley Blockpost (two here). The cattle wagons were scrapped in 1952, 30 and 31 being cut up at Hookagate.

Finally, there were three Inspection Trolleys; the make is not recorded, nor their fate; the W.D. record that a Fordson tractor was sent to Swindon in September 1951, along with the Merryweather manual fire engine.

DIMENSIONS OF S. & M.R. LOCOMOTIVES

No.	Name	Type	Cyls.	Wheel diam. Driving	Trailing	Wkg. press.	Wt. in. in w.o.
1.	*Gazelle*	0-4-2WT	4" x 9" inside	2' 3"	2' 3"		5t. 6c.
2.	*Severn*	0-4-2ST	14½" x 22" inside	4' 6"	3' 6½"	140	28t. 9c.
3.	*Hesperus*	0-6-0	16" x 20" inside	4' 7½"	—	160	47t. 7c.
5.	*Pyramus*	,,	,,	,,	,,	,,	,,
6.	*Thisbe*	,,	,,	,,	,,	,,	,,
4.	*Morous*	0-6-0ST	11" x 18" inside	3' 2"		,,	,,
5.	*Pyramus*	0-6-2T	14" x 22" outside	3' 6"	2' 6"	170	36t. 0c.
6.	*Thisbe*	,,	,,	,,	,,	,,	,,
7.	*Hecate*	0-6-0T	17" x 24" inside	4' 5½"	—	140	27t. 10c
8.	*Dido*	,,	,,	,,	,,	,,	,,
9.	*Daphne*	,,	,,	,,	,,	,,	,,
8108		0-6-0	17" x 24" inside	4' 3"	—	150	,
8182		,,	,,	,,	,,	,,	,,
8236		,,	,,	,,	,,	,,	,,

Military Control

Period 1941-1960

Reorganization of the line

IN July 1941, the author, in the course of a few days' stay at Shrewsbury, visited Kinnerley, the occasion calling for a three-mile walk from the nearest bus stop along leafy lanes whose age-old green made the war seem far away; as the objective was approached, however, military vehicles became increasingly frequent disturbers of the rural peace. Kinnerley yard was athrong with khaki, the sidings were full of military stock and dozens of the inevitable Nissen huts had sprung up, but the sentry with fixed bayonet dissuaded any lingering on the station bridge, for in those days interest in any railways—even of the most innocent kind—was apt to be frowned upon. It was clear that vital changes had come to the S. & M., but naturally it was some years later before any of the material facts were made generally available.

The W.D. requisitioned some 23 square miles of countryside in the vicinity of the railway for ammunition storage, and proceeded to lay out extensive series of sidings joining up numerous storage depots, each of these "yards" being connected with the S. & M. main line. Great changes took place in the appearance of the latter; most of the permanent way was relaid, in many places with 75 lb flat-bottomed track on concrete sleepers, and the whole was scrupulously cleared of weeds and ballasted with clinker instead of earth. The vastly increased volume of traffic anticipated called for re-organization of the layout at many points apart from the new "yards", the most interesting being the provision of an extensive marshalling yard at Hookagate near the site of the original Redhill station of "Potteries" days: the S. & M. Hookagate station disappeared in the process. A connection with the Welshpool line was put in and thereafter all exchange traffic was dealt with here instead of at Meole Brace, where the siding accommodation was very limited and the rather cramped position not suitable for expansion. A length of double track was laid at Ford and at certain other points adjoining the "yards", and a loop put in at Cruckton, though the latter was subsequently lifted; extra siding accommodation was provided at Llanymynech and three more sidings (one later removed) at Maesbrook. Five new halts—which for passenger purposes largely superseded the corresponding S. & M. stations—consisting of single concrete platforms with railings but no shelter, were built at Ford, Shrawardine, Pentre, Nesscliff and Kinnerley: Nesscliff lay beyond Nesscliff & Pentre station but the others were all a few hundred yards on the Shrewsbury side of the original stations. In addition, a short branch was constructed between Nesscliff and Edgerley, running north; this had connections with the main line running in both directions, being in effect an elongated reversing triangle At the terminus of the branch, a new four-platformed station, Lonsdale, was constructed. At Kinnerley itself, the layout was not much altered, but the track was relaid, complete new signalling installed and a new locomotive shed built, similar in size to the original and of the same type as those at Sheperdswell and Rolvenden.

The original signalling equipment had partially fallen into disuse* (the main signal at Kinnerley had indeed been blown off in a gale at Christmas 1938) and this was dismantled and replaced by a necessarily more elaborate system of upper quadrant semaphores for the most part, but lower quadrant signals at Hookagate, probably in deference to Great Western practice. New point installations were also supplied. The level crossings in general were not altered, an exception being Maesbrook, which was protected after the continental fashion by the "barbers' pole" type of gate, which is swung to

---

* In later years up to 1939, when only up platforms were used at Ford, etc., the up signal was lowered for down trains, to indicate to the driver that points were "correctly" set.

Birmingham Locomotive Club special, April 30th 1939. Gregory clears the track on the Criggion branch.
(H. W. Robinson)

the vertical to allow vehicles to pass; a keeper (in a sentry box) was maintained at this crossing.

Partly as a relic of S. & M. methods and partly to provide R.E. personnel with wider experience, a variety of operating methods were employed. From Shrewsbury to Hookagate East, Staff & Ticket was used, thence to Ford, Ticket & Telegraph, the "ticket" consisting of a special Army Form printed in three colours—white for authority to proceed in the normal way, orange to proceed with caution, and red to denote special caution. The next block section, Ford to Quarry (the west end of Ford "yard") was worked by the same method, but for the section from Quarry to Nesscliffe East a disc (from the Longmoor Military Rly!) was used instead of a ticket. The Miniature Staff operated Nesscliff East to Kinnerley, and from Kinnerley to Llanymynech the original S. & M. staff, used on the "one engine in steam" principle. From June 1st 1941 the W.D. operated all traffic on the main line, military and public.

Military requirements were very heavy during the war years, when a dozen locos were in steam daily; nevertheless, civilian traffic continued to be catered for by a daily goods between Shrewsbury and Llanymynech in the charge of Driver King (who had joined the S. & M. "temporarily" about 25 years earlier) and a guard; the latter sported a uniform lettered S. & M.R. in red and a "Railway Service" badge lettered similarly. Workmens' trains

The S. & M. track, as relaid by the War Department. View from Kinnerley station bridge, looking east, in October 1946. (*R. W. Clark*)

77

operated between Kinnerley and Llanymynech and to the "yards", and at weekends there were leave trains to Shrewsbury. After Victory day, some relaxation was permitted and civilian passengers were allowed to travel (free) on the mixed train which left Shrewsbury at 8.45 a.m., calling at all stations, including Lonsdale, to Kinnerley, though intermediate shunting was even more prolonged than in independant days. Responsibility for the working of the line is in the hands of the Operating Officer, a role filled by Mr. W. Proctor until that gentleman's untimely death in November 1947, when he was succeeded by Mr. C. H. Calder, M.B.E.: the staff totalled about 100.

The Criggion branch remained outside the scope of W.D. control and continued to exhibit the familiar features of the S. & M. of pre-war days— grass-grown track and thick hedges, deserted stations and the daily train of granite wagons to and from the Breidden quarries. In 1945, however, the weakening Melverley viaduct developed an alarming sag near the Criggion side; at first, extreme precautions were taken, the wagons being pushed up to and over the bridge by the B.Q.C.'s vertical-boilered "Sentinel" loco-motive, to be collected on the Kinnerley side by the S. & M. engine. Later, the "Sentinel" was allowed to cross the bridge at walking pace and take the train all the way to Kinnerley, the droppings from the chain drive leaving a continuous oily black line on the permanent way; unlike the two speed "Super Sentinel", this type was only intended for yard work and was not

Melverley viaduct, looking north. The piers of the original P.S. & N.W. viaduct can be seen in the river, close to the contractor's digger. May 14th 1948.　　(*E. S. Tonks*)

Shrawardine viaduct, showing sagging pillar of original viaduct, which was main reason for the rebuilding. June 26th 1955. (D. Clayton)

Shrawardine viaduct; a standard Royal Engineers design, replacing the original structure. The abutments are the original ones. April 15th 1959. (S. A. Leleux)

79

really suitable for main line use, having a top speed of about five m.p.h. In 1947, the decision made to replace the viaduct and some of preliminary work was undertaken by Great Western engineers; in the Spring of the following year, the Contractor, Messrs. A. E. Farr Ltd., of Reading, arrived and proceeded to clear the ground for the erection of the new viaduct upstream from the old, practically on the site of the original "Potteries" bridge. A 2′ 0″ line worked by a Lister petrol locomotive was laid down to carry the material for the earthworks, whilst the heavy equipment (wooden piles, etc.) were stacked on Melverley station platform.

To return to the main line: the W.D. continued to effect improvements in the equipment generally, in 1945 sinking a deep well by the locomotive shed on the side of the metals opposite to the old one, with a motor-driven pump for filling the large storage tank. Another engineering work of importance was the rebuilding by R.E. personnel of the up side of the Shrawardine viaduct, the original girders from this side being cut down and a Bailey Bridge structure capable of taking heavier loads placed on the pillars; whilst this work was proceeding, traffic continued to use the original S. & M. down track. The railway was not without its worries, however; in particular, the disastrous floods of the winter of 1947 took their toll, the river rising to the top of the embankment at Shrawardine, while at Maesbrook a bridge over a stream was completely swept away, stranding a "Dean" at Llanymynech. A minor but important improvement was the erection of seven miles of boundary fencing. A milk factory was built at Edgebold and a new point

Ex L.S.W.R. Royal saloon at Kinnerley, April 23rd 1939. (R. K. Cope)

installation put in in 1942, though the siding was laid with the original "Potteries" track! The siding was removed in 1954.

Locomotives and rolling stock

On requisition, the W.D. came into possession of five locomotives—*Gazelle*, *Hesperus* and the three "Colliers"—which, even had they been in first-class condition, were quite inadequate to cope with the work of reconstruction and the much heavier traffic, therefore steps were at once taken to augment the stock. Following S. & M. tradition, a decided preference for 0-6-0 tender engines was shown, and throughout the war years the bulk of the work on the railway was delegated to ex-G.W.R. "Deans". Very soon after the outbreak of war, the W.D. purchased 100 of this class (which had served well overseas in World War I); most of these were lost in France, but some never left the country and were retained for use at various home depots, together with a further eight borrowed in 1941 and taken into W.D. stock in 1942. The completion in the latter year of the S. & M. as a military establishment called for more motive power, which materialized in a variety that would have done credit to the late Colonel Stephens, but one by one most of the "oddments" were transferred elsewhere and replaced by more "Deans", which were well liked; in 1944, three American 0-6-0 T's were introduced, but only worked for a short time. Diesel traction was twice tried, but proved unsuitable. Power was also occasionally borrowed from

Abbey station on May 14th 1948. Rea bridge in foreground.   (*E. S. Tonks*)

81

Shrewsbury L.M.S. or G.W. depot.

The cessation of hostilities brought about a gradual reduction of traffic and batches of the hard-worked engines were withdrawn from service and sent to Hookagate yard for disposal or breaking up; but there seemed to be no hurry to get rid of them and the rows of rusting engines attracted many railway enthusiasts for a number of years. The rebuilding of Shrawardine viaduct brought about even greater changes, inasmuch as the axle-load limitation was thereby raised to permit the use of the standard "Austerity" 0-6-0 saddle tanks which, based on a design of the Hunslet Engine Co., had been ordered in large numbers by the W.D. from half a dozen builders. The arrival of these engines sounded the death-knoll of the "Deans" and by the middle of 1948, only one of the latter (No. 180) was at work. The changeover was by no means popular with the operating staff, however, as the new engines—though admirable for yard shunting—were not ideal for lengthy main line working, as the tanks required more frequent replenishment.

The livery of the "Deans" was plain black, latterly relieved in some cases by red lining; 70094 was named "Monty" in flowing characters flanked by (? laurel) leaves on the centre splasher; 70095 was embellished similarly with the letters R.E. flanking the Corps' crest. The first batch of "Austerity" saddles were camouflage green or brown, the later arrivals (from Cairnryan) black with yellow figures. The "Colliers", as already mentioned, were also camouflage green, though later they were given red lining. Most of the other engines were black, but there were some exceptions, e.g. 73 was green, and 72215 buff.

In addition to the steam stock, the W.D. introduced a fleet of petrol-driven inspection cars, supplied by Messrs. Wickham of Ware; these had two-cylinder JAP engines and friction drive, bodywork Spartan in character, and were very economical to run. The direction of travel was reversed simply by shifting the drive from one side of the flywheel to the other, but mechanical difficulties sometimes made it easier to turn the whole light car round bodily. The numbers of the cars are given in the table; at some time they bore simple figures 1, 2, etc., but these were changed from time to time.

When running, each car carried a flag, and was subject to some modifications of absolute block working, e.g. the down starting signal at Hookagate East was beyond a bridge, out of sight of the blockman, and a Wickham car approaching was allowed to proceed beyond the signal as far as the bridge. A steam engine was denied this privilege, because the blockman was made aware of its presence by a whistle or a column of smoke; Red Indian tactics beyond the petrol engine! One other internal combustion machine came to Kinnerley in September 1947; this was a Guy truck with flanged wheels and dubbed "The Harlech Express", and came from the W.D. rifle range rail-

The "Dean Goods" were understandably rarely photographed in their service with the W.D. and although the above illustration is not a S. & M. scene it merits inclusion as an example of their work. 70094 MONTY—later to work on the S. & M.—runs into Oxford with a forces leave train from Bicester W.D. Depot one Sunday afternoon, June 24th 1945.
(R. H. G. Simpson)

Withdrawn locomotives in store at Hookagate on May 18th 1948. Ex L.N.E.R. 0-6-0Ts, W.D. "Dean" 0-6-0s and ex S. & M. "Colliers".
(E. S. Tonks)

way on the sandhills at Harlech. 75191 was sent from Kinnerley to dismantle the latter line in 1948, as their diesel was out of commission.

For Passenger stock, the W.D. provided a mixed, though not remarkable, assortment—eight ex-L.T.S.R. centre corridor bogies, four ex-G.W.R. clerestory bogies, one ex-L.M.S. flat-topped bogie—all painted brown or camouflage green; class distinction took the form of "Officers only" or "A.T.S. only". Goods wagons came from far and wide—L.M.S., L.N.E., L. & Y., L.S.W., Melbourne Military Rly., etc., and included some new construction, e.g. one bogie sleeper, and a number of brakevans.

Listed below are the locomotives used on the S. & M.R. during the whole period of Military control; the table has been divided into sections for simplicity of reference.

1. Locomotives taken over from the S. & M.R., 1941

Number/name	Type	Cylinders	Builder	Maker's No.	Year Built
Gazelle	0-4-2WT	inside	Dodman	—	1893
to Longmoor for preservation, June 1950.					
Hesperus	0-6-0	inside	Beyer, Peacock	1517	1875
Scrapped Nov. 1941.					
8108 (2 until 1941)	0-6-0	inside	Crewe	1869	1874
8182	,,	,,	,,	2333	1879
8236	,,	,,	,,	2459	1881
All three locomotives to B.R.(W.R.) May 1950. Withdrawn July 1950 and scrapped at Swindon October 1950.					

2. Ex G.W.R. Dean 0-6-0 tender locomotives. (inside cylinders, built at Swindon)

Number/name	Maker's No.	Year Built
70093 (93 until 1944)	1363	1893
ex Melbourne Mil. Rly. 1943, formerly G.W.R. 2433; to East Kent Rly 1944, returned 1944. Sold 1948.		
70094 *Monty*	1199	1890
ex Bicester, Oxon, Dec. 1946; formerly G.W.R. 2399. Scrapped 1948.		
70095	1472	1896
ex Bicester, Oxon, 1946, formerly G.W.R. 2470. Scrapped Sept. 1948.		
70096 (96 until 1944, 2425 until 1942)	1295	1892
ex Longmoor Mil. Rly. 1941, formerly G.W.R. 2425; to Longmoor 1945, returned 1945. Scrapped Sept. 1948.		
70097 (97 until 1944, 2442 until 1942)	1372	1893
ex G.W.R. 2442, 1941. Sold 1948.		
70098 (98 until 1944, 2415 until 1942)	1285	1891
ex Bicester May 1942, formerly G.W.R. 2415. Scrapped Sept. 1948.		
70099 (99 until 1944)	1569	1897
ex Melbourne Mil. Rly. 1944, formerly G.W.R. 2528. Sold 1948.		
70169 (169 until 1944)	1481	1896
ex Kineton Depot, Warwicks., 1948, formerly G.W.R. 2479. Scrapped or sold about 1948.		
70170 (170 until 1944)	1577	1897
ex Kineton Depot, Mar. 1944, formerly G.W.R. 2536. To Cairnryan, Stranraer, 1945.		

Nesscliff Halt; one of several built to serve military depots. Wickham railcar in foreground. May 14th 1948. (E. S. Tonks)

Kinnerley shed yard, in May 1948. Wickham car shed on left; main shed with new (left) and old (right) water towers. (E. S. Tonks)

70175	(175 until 1944)			1552	1897

70175 (175 until 1944)   1552   1897
ex G.W.R. 2511, 1942. Scrapped about Aug. 1948.

176   1639   1897
ex G.W.R. 2558, 1942. Scrapped Jan. 1944.

70180   1555   1897
ex ?, Kent, 1946, formerly G.W.R. 2514. Scrapped or sold.

70196   1657   1898
ex ?, 1946, formerly G.W.R. 2576. Scrapped or sold 1948.

70197   1581   1897
ex East Kent Railway 1946, formerly G.W.R. 2540. Scrapped about Sept. 1948.

200   1633   1897
ex Longmoor, 1941, formerly G.W.R. 2552. To Stonar, Kent, 1943.

NOTE: The last four locomotives were fitted with additional pannier tanks.

3. Miscellaneous W.D. Locomotives

Number/name	Type	Cylinders	Builder	Maker's No.	Year Built
73	0-6-0PT	outside	W. G. Bagnall	2643	1941

73 0-6-0PT outside W. G. Bagnall 2643 1941
New to S. & M. To Sinfin Lane Depot, Derby, 1944.

70084 0-6-0T inside Stratford — 1895
ex Old Dalby Depot, Notts., 1945, formerly L.N.E.R. 7388. To John Lysaght Ltd., Scunthorpe, May 1948.

70091 0-6-0T inside Stratford — 1904
ex Cairnryan, Stranraer, 1944, formerly L.N.E.R. 7088. To John Lysaght Ltd., Scunthorpe, May 1948.

92 (formerly 654) 0-6-0ST inside Manning Wardle 654 1877
ex Long Marston Depot, Warwicks., April 1942, formerly Oxford Greystone Lime Co. Ltd., Surrey. To Sinfin Lane, May 1945.

202 *Tartar* 0-4-0ST outside Avonside 1407 1899
ex Queensferry Depot 1942. To Queensferry 1943.

212 0-6-0 inside Stratford — 1889
ex London Film Productions Ltd., 1942, formerly L.N.E.R. 7835. Scrapped 1944.

221 0-6-0 inside Stratford — 1888
ex London Film Productions Ltd., 1942, formerly L.N.E.R. 7541. Scrapped 1944.

71872 *Ashford* 0-6-0ST outside Avonside 1872 1920
(1872 until 1944)
ex G. Cohen Sons & Co. Ltd., 1942. To Abelson & Co. (Engineers) Ltd., Birmingham, Mar. 1949.

4. Main line locomotives on loan

Number	Type	Cylinders	Builder	Maker's No.	Year Built
982	0-4-0T	inside	Darlington	—	1923

982 0-4-0T inside Darlington — 1923
ex L.N.E.R. 982, 1942. Returned 1943.

3014 0-4-0WT inside Crewe 2218 1880
ex L.M.S.R., Crewe, 1942. To A. R. Adams Ltd., Newport, Mon., 1943.

3015 0-4-0WT inside Crewe 2511 1882
ex L.M.S.R., Crewe, 1942. To A. R. Adams Ltd., Newport, Mon., 1943.

5. Ex U.S.A. Transportation Corps. (outside cylinder 0-6-0T locomotives)

Number	Builder	Maker's No.	Year Built
1395	H. K. Porter, U.S.A.	7509	1942
1399	,,	7513	1942
1427	,,	7541	1943

All acquired from an unknown source in 1944 and disosed of similarly in 1945.

Llanymynech Junction; S. & M. left, Welshpool line centre, Llanfyllin line right.
June 15th 1948.   (*P. J. Garland*)

6. Standard W.D. "Austerity" tank locomotives (inside cylinder 0-6-0ST)

Number	Builder	Maker's No.	Year Built
75171	W. G. Bagnall	2759	1944

ex Donnington Depot, Shrops., Sept. 1947; to Bramley Depot, Hants., *c.* 1949.

75193	R. Stephenson Hawthorn	7143	1944

ex Bicester, Oxon, 1949, to Bicester *c.* Nov. 1950.

103 (formerly 75036)	Hunslet	2885	1943

ex Hunslet Engine Co. Ltd., 1949, formerly at Sudbury Depot, Staffs.; to Long Marston Depot, Warwicks. by June 1954.

120 (formerly 71528)	Barclay	2182	1944

ex Donnington, c. 1950; returned to Donnington by Nov. 1953.

121	Hunslet	2894	1943

ex Hunslet Engine Co. Ltd., by April 1954, formerly at Arncott Depot; to Bicester during period Sept. 1956-Dec. 1957.

124 (formerly 75049)	Hunslet	2898	1943

ex Cairnryan, Stranraer, 1948; to Hunslet Engine Co. Ltd., *c.* 1953.

125	R. Stephenson Hawthorn	7099	1943

ex Bicester, Feb. 1957; returned Mar. 1960.

130	Hunslet	3157	1944

ex Tidworth Depot, Hants., by Feb. 1954. To Bicester Dec. 1954.

135	Hunslet	3171	1944

ex Hunslet Engine Co. Ltd., *c.* Oct. 1951, formerly at Sudbury; to Long Marston *c.* Jan. 1953.

139 (formerly 75141)	Hunslet	3192	1944

ex Old Dalby Depot, Notts., Jan. 1947 to Lockinge Depot, Berks. by Jan 1952.

141	Hunslet	3195	1944

ex Donnington, during period 1954-June 1955, to Bicester Mar. 1960.

143 (formerly 75152)	W. G. Bagnall	2740	1944

ex Cairnryan, Feb. 1947; to Donnington *c.* Nov. 1953. Ex W. G. Bagnall Ltd., *c.* July 1956; to Bicester Mar. 1960.

146 (formerly 75165)   W. G. Bagnall   2753   1944
ex Bicester *c.* 1948; to Yorkshire Engine Co. Ltd., during period Aug. 1950-Oct. 1951, later at Long Marston.
151 (formerly 75187)   R. Stephenson Hawthorn   7137   1944
ex Old Dalby, Jan. 1947. To Bicester *c.* 1953.
153 (formerly 75191)   R. Stephenson Hawthorn   7141   1944
ex Cairnryan Feb. 1947; to Hunslet Engine Co. Ltd., 1954, later at Donnington.
154 (formerly 75192)   R. Stephenson Hawthorn   7142   1944
ex Donnington by Aug. 1950: to Yorkshire Engine Co. Ltd., by Sept. 1954, later at Longtown Depot, near Carlisle.
167 (formerly 71531)   Barclay   2185   1945
ex Cairnryan 1948; to Arncott during period June 1955-June 1956.
185   Vulcan Foundry   5276   1945
ex ?, *c.* 1952, formerly at Kineton Depot, Warwicks.; to Bicester Aug. 1957.
188   Vulcan Foundry   5284   1945
ex Arncott May 1957; to Bicester Mar. 1960.
189   Vulcan Foundry   5285   1945
ex Yorkshire Engine Co. Ltd., by Feb. 1954, formerly at Lockinge; to Bicester *c.* Oct. 1956.
193   Hunslet   3793   1953
ex Woolmer Store, Hants, by Feb. 1955: to Bicester May 1960.

7. W. D. diesel locomotives

No.	Type	Builder	Maker's No.	Year Built	Origin and Disposal
32	0-4-0DM	Drewry	2159	1941	new; to Chilwell Depot, Notts, *c.* 1942
40	,,	Barclay	355	1941	new; to Longmoor, Dec. 1943
44	,,	,,	359	194ᴉ	new; to Queensferry, 1943
45	,,	,,	360	1942	new; to Queensferry, 1943
48	,,	,,	363	1942	new; to Longmoor, Dec. 1943
72215	4wDM	Ruston Hornsby	224347	1945	new; to Longmoor, 1945
8201	0-4-0DH	North British	27422	1955	new; to ?, at Tidworth by Aug. 1959

8. Petrol railcars (all 4wPM)

No.		Builder	Maker's No.	Origin and Disposal
2941	No. 1	Wickham	2941	? Scrapped or sold
2942		,,	2942	,,
2943	2	,,	2943	,,
3376		,,	3376	,,
3410		,,	3410	,,
11965		,,	3373	,,
11966		,,	3374 (?)	,,
11967		,,	3375	,,
9004*		,,		ex ? by Mar. 1958; Scrapped or sold
9005*		,,		ex ? by Mar. 1958; Scrapped or sold
9006*		,,		ex ? by July 1956; Scrapped or sold
9007*		,,		ex ? by Mar. 1958; Scrapped or sold
9103		Drewry	2323	ex ? by Mar. 1958; to Bramley by July 1959
9104		,,	2324	ex ? by Mar. 1958; to Bicester by Sept. 1960
9105		,,	2325	ex ? by July 1956; to Bicester by Sept. 1960
9106		,,	2326	ex ? by July 1956; to Bicester by Aug. 1959
A 14620		Guy (reb. of Guy road truck)		ex Harlech Depot by Sept. 1947; scrapped or sold.

* These railcars may, in fact, be the Wickham cars listed above with different numbers.

Birmingham Locomotive Club special on June 26th 1955 in Ford yard with W.D. loco 193. The first two coaches were built for the L.T.&S.R. Ealing — Southend through Service.   *(T. J. Edgington)*

Abbey Station in October 1959. War Department saddle tank 125.   *(J. A. Peden)*

The Post-War Period

The war years were, of course, the busiest that the S. & M. had ever seen, and in the period 1941-5 over a million tons of military traffic was moved. Civilian traffic increased slightly over pre-war years; in 1939 it was about 35,000 tons, while for 1942-7 the tonnage was over 40,000 per year. The increase was probably due to the better service offered by the W.D. and to the wartime policy of utilizing rail facilities in preference to road.

The cessation of hostilities naturally brought about a rapid reduction in the volume of military traffic, and some of the yards were closed and the tracks lifted; in 1947 control of the line was transferred from W.D. Military to W.D. Civilian status, and Kinnerley was established as Headquarters of one of the Home Railways groups of the Royal Engineers. The S. & M. itself was one of the smaller railways embraced by the Transport Act, 1947, a fact announced by the appearance on January 1st 1948 of the "Railway Executive" poster on Abbey Station, and in May of that year B.R. stock was distributed to S. & M. shareholders for the current market value of the stock —6d. per £10 share, or £25 for the whole railway. The Shropshire Railways Co. was also taken over as a "paper" transfer, as the Company did not operate any lines; but they continued to receive the £886 annually for the section of line ceded to the Cambrian Railways in 1900. In fact, the Railway Executive, Western Region, only exercised direct control over the Criggion branch, and took part in the rebuilding of Melverley Viaduct, as we have seen; and in May 1948 a group of half a dozen men were hard at work at Chapel Lane, clearing of weeds a siding that had probably not been used for twenty years. . . . A crop of "Trespassers W . . ." notices appeared with slips of paper pasted over the "H. F. Stephens, General Manager" — evidently the S. & M. had a large stock of these!

It is perhaps surprising that the S. & M. continued to survive; that it did so was of course entirely conditional on W.D. occupation, and while this was in force the stone traffic from Criggion was maintained, with the inevitable tendency to gradually diminish. The military depots were closed one by one, but it was a slow process; meanwhile the W.D. undertook a cleaning-up operation. The bulk of the surplus locomotives were disposed of in 1948, as already recorded, and in later years other items found their way to Swindon, never to return. Military control had destroyed much of the rustic charm of the old S. & M. but it had saved the line from extinction and there were still a lot of things to interest the enthusiasts, who were encouraged by genial management personified by Mr. C. H. Calder. Enthusiast "specials" were run from time to time; a train hauled by an "Austerity" 0-6-0ST would pick up the party at Abbey station and proceed

1970 view of Llandrinio Road station, with Breidden Hills in background.
(E. C. Griffith)

Criggion station, looking north, in 1970.   *(E. C. Griffith)*

to Kinnerley, with stops at points of interest en route, the chief naturally being Shrawardine viaduct; here the passengers would alight and the train would then be shunted onto the viaduct to give an interesting shot for the cameramen. At Kinnerley, exploration of the Criggion branch was made possible by using Wickham railcars, as the steam loco was too heavy. In this way on June 26th 1955 the Birmingham Locomotive Club revived memories of its historic trip behind *Gazelle* sixteen years and one war earlier. Like the loco and carriages, the main line had changed for the better very considerably — but the Criggion branch seemed to have altered hardly at all — the same grass grown track, even the straying cows still. So enthusiasts sampled the nostalgic delights of the S. & M.

Abbey station in 1970 showing present B.R. oil tanker sidings. Further si dings are on left of crane.
(E. C. Griffith)

Closure and Dismantlement

"So the enthusiasts sampled the nostalgic delights of the S. & M." we said at the end of the last chapter; but it was obvious that the end could not be far away, and in 1959 the War Department decided to terminate its use of the railway. The last military depots were closed and the tracks serving them lifted. The Criggion branch was closed completely in December 1959 after the cessation of the quarry traffic, and civilian traffic on the main line officially ceased from February 29th 1960; actually the last scheduled train worked from Abbey on February 26th 1960, when work commenced immediately on putting in a connection from the British Railway's Severn Valley line to Abbey Goods Yard, a portion of which B.R. intended to retain. From this date, Llanymynech became the outlet for military traffic being removed from the line. The Stephenson Locomotive Society ran a farewell trip on March 20th 1960. On March 31st 1960, all military traffic having been removed, the line was formally handed over to B.R.(W.R.), for dismantlement.

The Criggion branch, which the W.D. had never requisitioned, was taken up first; the main line was subject to some financial adjustments between the W.D. and the B.T.C.; for example, the B.T.C. agreed to pay the W.D. the difference in scrap value between the new track and the original S. & M. track. When these details were settled, dismantling of the main line took place, in 1962, the contractors being Messrs. Marple & Gillott Ltd. of Sheffield, who used a diesel locomotive for the track lifting.

Shrawardine viaduct was dismantled in 1962; but Melverley viaduct, on the already lifted Criggion branch, was reconstructed in May 1962 as a road bridge, and the trackbed between Crew Green crossing and the level crossing south-south-east of Melverley church made up as a minor road. There was no other crossing of the river between Montford Bridge and the one at Llandrinio, and the new bridge is a boon to motorists with business in the extremely scattered village of Melverley.

When the tracklifting was completed, the only remaining section of track was the siding in Abbey station yard serving the petrol depot and connected to the truncated Severn Valley line; the depot is serviced by a B.R. 0-6-0DE shunter. The future of this section is in doubt; the *Liverpool Daily Post* of November 29th 1969 suggested that it was "likely to be closed soon" but it was still in operation in mid-1971.

The B.R. line through Llanymynech was closed to passengers from Monday January 18th 1965, the last train running on January 16th 1965. On this date the Welshpool-Oswestry line was closed to passengers, the Buttington Junction-Llynclys Junction section (including Llanymynech) closed completely; but the Oswestry-Llynclys section is still open for use by trains serving Nantmawr quarries (which is where the "Potts" really started!). Hookagate Exchange Sidings became a Rail Welding Depot, served formerly by a Ruston & Hornsby 0-6-0DE, PWM 654, but more recently by a standard 0-6-0DE, D3510.

In spite of all this, Kinnerley is not devoid of locomotive interest; the Welsh Highland Railway Society, formed to revive and operate portions of the Welsh Highland Railway, were precluded by legal difficulties from getting possession of the trackbed. They had meanwhile acquired a number of locomotives, some track and other equipment, and all this was stored at Kinnerley through the generosity of the landowner. Most of the 2' 0" gauge locos are housed in a long black Nissen type shed in the V between the Llanymynech and Criggion lines and outside track is stacked, also two locos of wider gauge. Behind the black shed is the lorry formerly used for carrying track etc. up to Beddgelert; as negotiations have been protracted, this lorry is now derelict. One other 2' 0" gauge loco, in a dismantled condition, is housed in a brick building on the other side of the Criggion branch. The selection of this site seems extraordinary, even bearing in mind that the Company's office is at Shrewsbury; the stock could have been housed anywhere where space was available — but the Company's officers are "railway orientated" and Kinnerley must have exerted a special appeal. The owners are now the Welsh Hignland Light Railway (1964) Ltd.

Conclusion

The history of the S. & M. and its predecessors spans a hundred years and there can be few railways that have had so many changes in title and fortune. The track has all gone apart from the siding at Shrewsbury, but the route is clear (though not always passable) almost throughout, and many buildings and other relics survive. These include remains from each period of the railway's chequered history, and the enthusiast visiting the sites will find it a fascinating study, placing the remains in their respective periods. Most of brick station buildings from "Potts" days survive, and there are bridges reconstructed by the Shropshire Railways (e.g. west of Hanwood Road). Station platforms remain, sometimes with wagon bodies on them, mostly of W.D. origin. There are plenty of traces of the wartime military depots served by the line. Some of the bridges have gone, for example the one on the A 458 at Ford; but others remain, as at Kinnerley. Shrawardine viaduct has gone, too, alas and Melverley rebuilt as a road bridge. For those without their own transport, Vaggs Motors provide a service, including one from Shrewsbury to Llanymynech, hopping from one side of the railway route to the other.

The Criggion branch never had the complete clean-up that the W.D. applied to the main line and, as no earthworks were necessary in this flat area of the Severn plain, it has been absorbed more completely into the surrounding countryside. Nevertheless, the hedges remain and level crossings can be identified; Melverley station platform and building survive, but the unique many-arched bridge has gone and the road stands only a foot or

Pentre station, looking east, in 1970. W.D. van on platform.　　　　(*E. C. Griffith*)

Kinnerley station now, looking west. The station is in the foreground; locomotive shed (now a contractor's premises) to the left of the poplar. The Llanymynech line is upper right and the Welsh Highland stock to left of this.　　　　(*E. C. Griffith*)

95

Melverley viaduct—final form as a road bridge. (E. C. Griffith)

Llanymynech station in 1970, showing remains of B.R. station in centre, with curved S. & M. platform to the extreme right. Llanymynech Hill quarry (where the P.S. & N.W. started), top left. (E. C. Griffith)

two above the former rail level. At Llanymynech the main line station has disappeared as well as the S. & M., though the platforms remain, and a factory area has sprung up behind them. We have not particularized the remains at every point of the route; for one thing, they will change with every passing year, and further, a catalogue merely helps those who collect rather than preserve, to strip the railway of the more saleable relics. Those who explore the hard way are more entitled to them, we feel.

Kinnerley will always exert the greatest nostalgic appeal to the enthusiast and there is plenty to see here — and plenty to muse on, too. There is the station, the bridge carrying the road over it, from which one can see the main line stretching away each side; the locomotive shed, currently owned by a plant-hire contractor, still stands, whilst the loco enthusiast is catered for by the Welsh Highland Railway stock. Indeed, one is tempted to believe the Colonel's spirit could rest in peace here still.

Outside the premises, few relics survive; *Gazelle* is in the care of the Ministry of Defence and there seems no fear of her being scrapped. No other locos remain, but the author has a nameplate of *Thisbe* (bought for half a crown from the scrapman on the site), and Vic Bradley has a cast iron Shropshire Railways plate. Tickets, handbills and the like are in various private collections and photographs in fair number.

The feature of the countryside that governed the life of the "Potts" and the Shropshire & Montgomeryshire in its various guises was the sparseness of the population. It ruined the "Potts", stifled the Shropshire Railways, and almost killed Colonel Stephens' S. & M.; on the other hand it was precisely this feature that attracted the attention of the War Department; and, now that the railway has gone, it will, we hope, ensure that the remains be undisturbed for a long, long time.

Exploration of the Shropshire Railways: T. R. Perkins party in cutting near Shrawardine Viaduct, June 29th 1903. Left to right—Reeves, G. M. Perkins, F. E. Fox-Davies, T. R. Perkins, Morton.   *(F. E. Fox-Davies)*

Exploration of the Shropshire & Montgomeryshire Railway: Birmingham Locomotive Club party at Criggion, April 23rd 1939. Top Row—R. J. Buckley, J. K. Leary, S. Checketts, R. Alford, E. L. McKears, W. P. Edmunds (behind funnel), E. J. Jones, F. G. Alford, A. Holland, T. Shearer, R. K. Cope, A. W. Croughton, N. W. Wilks, E. S. Tonks, Bottom Row—J. K. W. Steer, R. Tedstone, C. A. Lowe, B. J. Drew, A. N. H. Glover, L. W. Perkins, K. C. Bridge, F. H. Latham, H. G. Tonks.   *(A. N. H. Glover)*

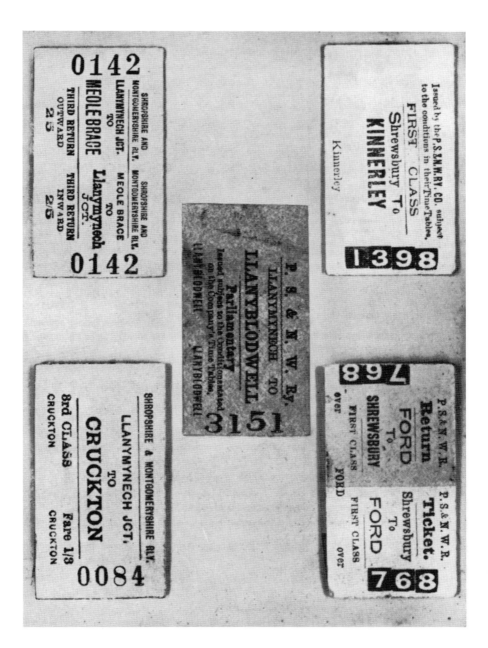

Specimen tickets of the P.S. & N.W.R. and S. & M. Railways.

APPENDIX I Locomotives which were operated at Ceiriog Quarries, Criggion. The successive owners of the quarries were Pyx Granite Co. Ltd; Granham's Moor Quarries Co. Ltd. (from 1913); Ceiriog Granite Co. Ltd. (from about 1925) and British Quarrying Co. Ltd. (from February 4th 1929).

Gauge: 4' 8½"

Number/name	Type	Cylinders	Builder	Maker's No.	Year Built
—	0-4-0ST	outside	Baldwin, U.S.A.	45335	1917

ex Railway Operating Dept., scrapped c. 1928.

| — | 4wVBT | vertical;
(geared drive) | Sentinel | 7026 | 1927 |

New to Ceiriog. Scrapped on site by James Rollason of Wellington, Apr. 1962.

The quarry line made an end-on connection with S. & M. at the Criggion terminus; crossing the lane there to it then ran in the same south westerly direction, along the foot of the steep escarpment of Breidden Hill, for nearly a mile. The quarry crushing plant was about half way along this stretch of line and at this point there were sidings for the reception of wagons awaiting filling at the crusher. The locomotive shed lay on the Criggion side of the crusher. Track was of simple quarry type with spiked rails on sleepers in earth ballast — lower than main line standards naturally but not far short of those of the S. & M.! Rail traffic from the quarries ceased in December 1959.

Gauge: 2' 0"

| *Jack* | 0-4-0ST | outside | W. G. Bagnall | 1650 | 1901 |

ex Cliffe Hill Granite Co. Ltd., Leics., c. 1914. Scrapped c. 1930.

The narrow gauge line ran between the quarry and the standard gauge sidings. This tramway was replaced by road transport about 1930.

APPENDIX II Locomotives owned by The Welsh Highland Light Railway (1964) Ltd. and stored at Kinnerley.

Gauge: 3' 0"

| *Handyman* | 0-4-0ST | outside | Hudswell Clarke | 573 | 1900 |

ex Staveley Minerals Ltd., Scaldwell Quarries, Northants., May 1964. Stored on Welshpool & Llanfair Railway property until April 1969.

Gauge: 2' 6"

| — | 0-6-0DM | | J. Fowler | 4160004 | 1951 |

ex A.P.C.M. Ltd., Lower Penarth Cement Works, Nov. 1969.

Gauge: 2' 0"

| *Russell* | 2-6-2T | outside | Hunslet | 901 | 1906 |

New to North Wales Narrow Gauge Railway (later known as Welsh Highland Railway). Preserved by Birmingham Locomotive Club and stored at Towyn until presented to W.H.L.R in April 1965, and moved to Kinnerley. To Hunslet Engine Co. Ltd., for new boiler, Nov. 1969-Feb. 1971.

| — | 0-6-0T | outside | Barclay | 1578 | 1918 |

ex Stewarts & Lloyds Ltd., Bilston Furnaces, Staffs. Preserved by N. Melhuish and presented to W.H.L.R. in October 1963.

| — | 4wPM | | Motor Rail | 6031 | 1936 |

ex Johnston Bros. (Salop) Ltd., Ercall Quarry, Wellington, Feb. 1963. Stored elsewhere until May 1965.

British Quarrying Co. Ltd., Criggion. Sentinel 7026 in its shed.

(Authors collection)

S. & M. horse drawn manual fire-engine. Swindon dump, November 11th 1951.

(T. J. Edgington)

| 6 | 4wDM | Ruston Hornsby | 297030 | 1952 |
| — | ,, | ,, | 354068 | 1953 |

Both ex Blockleys Ltd., Hadley Lodge Brickworks, Mar. 1963. Stored elsewhere until Oct. 1964.

| — | 4wDM | ,, | 327904 | 1951 |

ex Wenlock Stone & Concrete Products Ltd., Much Wenlock, 1966.
